SEMA SIMAI Springer Series

ICIAM 2019 SEMA SIMAI Springer Series

Volume 10

This sub-series of the SEMA SIMAI Springer Series aims to publish some of the most relevant results presented at the ICIAM 2019 conference held in Valencia in July 2019.

The sub-series is managed by an independent Editorial Board, and will include peer-reviewed content only, including the Invited Speakers volume as well as books resulting from mini-symposia and collateral workshops.

The series is aimed at providing useful reference material to academic and researchers at an international level.

More information about this subseries at http://www.springer.com/series/16499

Patrizia Donato • Manuel Luna-Laynez
Editors

Emerging Problems in the Homogenization of Partial Differential Equations

 Springer

Editors
Patrizia Donato
Laboratoire de Mathématiques Raphaël
Salem UMR6085 CNRS-UR
Université de Rouen Normandie
Saint Étienne du Rouvray, France

Manuel Luna-Laynez
Departamento de Ecuaciones Diferenciales
y Análisis Numérico
Universidad de Sevilla
Sevilla, Spain

ISSN 2199-3041 ISSN 2199-305X (electronic)
SEMA SIMAI Springer Series
ISBN 978-3-030-62032-5 ISBN 978-3-030-62030-1 (eBook)
https://doi.org/10.1007/978-3-030-62030-1

This Springer imprint is published by the registered company Springer Nature Switzerland AG.
The registered company address is: Gewerbestrasse 11, 6330 Cham, Switzerland

Preface

This book contains some of the results presented at the minisymposium titled *Emerging Problems in the Homogenization of Partial Differential Equations*, held during the ICIAM2019 conference in Valencia in July 2019. The quality of the works presented, as well as the successful reception of the minisymposium among those attending its sessions, led its organizers to accept Springer's invitation to publish this volume.

In order to introduce nonspecialist readers to the subject, let us mention that the aim of the mathematical homogenization is to model microscopically heterogeneous media, providing macroscopic models that describe their effective behavior. As an example of these media, we can consider composite materials, which are characterized by the fact that they contain two or more finely mixed constituents. These materials are widely used nowadays in industries, due to their properties. Indeed, they have in general a better behavior than the average behavior of its constituents. Well-known examples are the multifilamentary superconducting composites that are used in the composition of optical fibers.

Generally speaking, in a composite the heterogeneities are small compared with its global dimension. So, two scales characterize the material: the microscopic one, describing the heterogeneities, and the macroscopic one, describing the global behavior of the composite. From the macroscopic point of view, the composite looks like a *homogeneous* material. The aim of homogenization is precisely to give the macroscopic properties of the composite by taking into account the properties of the microscopic structure. From the mathematical point of view, this leads to study the asymptotic behavior of a Partial Differential Equation (PDE) depending on a small parameter, here denoted by ε, as ε tends to zero. The parameter describes the heterogeneities, which are "small" compared with the global size of the material. The main questions arising in this asymptotic analysis are:

- Does the solution of the PDE converge to some limit function?
- If that is true, does the limit function solve some limit boundary value problem and can we describe it explicitly?

Answering these questions is the aim of the mathematical theory of homogenization. Let us point out that since the coefficients of the PDE describe the characteristic of the material at the microscale, it is not realistic to suppose that the coefficients are smooth, for instance, continuous. Consequently, in general, the suitable framework is that of weak solutions in Sobolev spaces and variational formulations. Hence, the main challenge when passing to the limit is how to deal with products of two (or more) weakly convergent functions, which, as well know, do not converge to the product of their weak limit.

Perhaps the most classic homogenization problem appears when we study thermal properties of composite materials whose constituents are periodically distributed, and we have to pass to the limit in a second-order PDE with ε-periodic rapidly oscillating coefficients describing the stationary heat diffusion. When the heterogeneities are periodically distributed, the limit problem, the so-called homogenized problem, has constant coefficients, which can be described explicitly. Somehow this means that for small values of the parameter ε, the behavior of the composite material can be properly approximated by the behavior of another simpler material (a homogeneous material). Replacing the oscillating problem by the homogenized one allows to analyze the properties of the composite material in a much less complicated way, and in particular, to make easier numerical simulations. These classic results for the periodic setting have been later generalized to random composite material (i.e., the constituents are distributed in the medium according to a certain statistical law) and even to arbitrarily heterogeneous materials (which are neither periodically nor statistically homogeneous).

Composite materials are simply a kind of a wide variety of microscopically heterogeneous media, which can be studied by means of the mathematical homogenization theory. Others appear when we consider materials with holes and/or oscillating boundaries. This is, for example, the case of reticulated structures, made of thin beams or plates periodically distributed, which are very common in engineering and architecture. The study of these materials leads to consider PDE posed in perforated domains, with holes of size $\delta(\varepsilon)$ and distributed with ε-periodicity in each axis direction (here, $\delta(\varepsilon)$ is another positive parameter smaller than ε). In this setting, additional difficulties arise, since the PDE is posed in a varying domain and then its solution belongs to a Sobolev space that also depends on the small parameter ε. Remark that these homogenization problems play a main key in the optimal design of materials, when studying the limit behavior of minimizing sequences in order to prove the existence of an optimal shape or to obtain a relaxed formulation.

Although the first applications of homogenization come from engineering, its use is increasingly frequent in other fields. This is due to the fact that the theory constitutes a powerful mathematical tool for the study of complex systems that presents several elements (characteristics, constituents, etc.) of very different scales, very heterogeneous, and homogenized models provide an effective (macroscopic) description of the system that takes into account the influence of the different scales. Consequently, nowadays, the homogenization theory is used actually almost in any discipline, that is, the traditional ones such as physics, engineering, mechanics, chemistry, and economy but also biology and medicine. Let us highlight the case of the health sciences, where the systems under study are so highly complex that it is extremely difficult to study them using models that collect all scales (e.g., the blood flow in vessels, the transfer of oxygen from the alveoli of the lungs to the blood and of carbon dioxide conversely, and the skin pores among other).

The mathematical theory of homogenization has been well founded and widely developed from different approaches in the last decades. Once wide performance methods were well established, more and more researchers have expressed a high interest in them and in their application to new challenging problems, dealing with more realistic and complex models, as well as with more difficult mathematical problems. These new challenges, in turn, are an inspiration for method's improvement and development.

The aim of the minisymposium *Emerging Problems in the Homogenization of Partial Differential Equations* was to put together renowned specialists from all over the world overcovering a wide range of emerging challenging problems in the field. We have been very fortunate that our invitations to give a talk were accepted by so many well-known mathematicians, working in so many different countries and on a large spectrum of areas and problems and using many different methods. In gratitude, we recall below the list of authors of all the communications presented (we highlight the speakers in bold):

- Marc Briane (INSA de Rennes, France), **Juan Casado-Díaz** (Universidad de Sevilla, Spain).
- **Elisa Davoli** (University of Vienna, Austria), Irene Fonseca (Carnegie Mellon University, USA).
- **Daniela Giachetti** (Università di Roma La Sapienza, Italy).
- Antonio Gaudiello (Universitá Degli Studi Di Napoli Federico II, Italy), **Olivier Guibé** (Université de Rouen Normandie, France), Francois Murat (Laboratoire Jacques-Louis Lions, France).
- **María Eugenia Pérez-Martínez** (Universidad de Cantabria, Spain).
- Carlos Jerez-Hanckes (Universidad Adolfo Ibáñez, Chile), Isabel A. Martínez (Pontificia Universidad Católica de Chile, Chile), **Irina Pettersson** (University of Gävle, Sweden), Volodymyr Rybalko (Institute For Low Temperature Physics and Engineering, Ukraine).
- **Ben Schweizer** (TU Dortmund, Germany).
- **Grigor Nika** (Weierstrass Institute for Applied Analysis and Stochastics, Germany), Bogdan Vernescu (Worcester Polytechnic Institute, USA).

The communications covered a large range of topics, including the influence of a strongly oscillating magnetic field in an elastic body, new microstructure of materials exhibiting interesting, and technologically powerful, elastic and magnetic behaviors, relaxation and homogenization in the framework of A-quasiconvexity for differential operators with coefficients depending on the space variable, problems with weak regularity data involving renormalized solutions, singular nonlinear problems, eigenvalue problems for complicated shapes of the domain, homogenization of partial differential problems with strongly alternating boundary conditions of

Robin type with large parameters, multiscale analysis of the potential action along a neuron with a myelinated axon, dispersive long-time behavior of wave propagation, and multiscale model of magnetorheological suspensions.

It is our desire to conclude this Preface by showing in a special way our thanks to all the authors who agreed to participate in the publication of this volume.

Rouen, France   Patrizia Donato
Sevilla, Spain   Manuel Luna-Laynez

Contents

About the Editors

Patrizia Donato, Professor Emeritus at the University of Rouen Normandie, is author of 3 books and about 100 international articles on Partial Differential Equations, in particular their Homogenization. She has given about 100 lectures and seminars and 15 research courses in several countries, and organized about 20 international scientific events. She directed or co-directed 15 PhD theses. Member of the EMS Ethics Committee, and of the editorial board of Asymptotic Analysis, and Ricerche di Matematica.

Manuel Luna-Laynez is Professor at the University of Seville. He is currently the Director of the Department of Differential Equations and Numerical Analysis. His main research interests concern homogenization theory and asymptotic analysis of partial differential equations posed in thin domains. He is the author of numerous publications focused on applications to the optimal design of materials, behavior of fluids in domains with rough boundaries, porous media, and elastic multistructures.

Micro-geometry Effects on the Nonlinear Effective Yield Strength Response of Magnetorheological Fluids

Grigor Nika and Bogdan Vernescu

Abstract We use the novel constitutive model in Nika and Vernescu (Z Angew Math Phys 71:1–19, 2020), derived using the homogenization method, to investigate the effect particle chain microstructures have on the properties of the magnetorheological fluid. The model allows to compute the constitutive coefficients for different geometries. Different geometrical realizations of chains can significantly alter the magnetorheological effect of the suspension. Numerical simulations suggest that particle size is also important as the increase in the overall particle surface area can lead to a decrease in the overall magnetorheological effect while keeping the volume fraction constant.

1 Introduction

Magnetorheological (MR) fluids are a suspension of non-colloidal, ferromagnetic micron-sized particles in a non-magnetizable carrier fluid. They were discovered by J. Rabinow in 1948 [17]. The ability of magnetorheological fluids [17] to transform from a liquid to a semi-solid state in a matter of milliseconds makes them desirable for many applications [5, 12]. MR fluids are part of a larger class of suspensions of rigid particles, known as *smart* materials, for which their rheological properties can be controlled by the interaction with a magnetic or an electric field. Hence, accurate models for numerical simulations constructed on a sound analytical basis are required.

G. Nika (✉)
Weierstrass Institute for Applied Analysis and Stochastics, Berlin, Germany
e-mail: grigor.nika@wias-berlin.de

B. Vernescu
Mathematical Sciences, Worcester Polytechnic Institute, Worcester, MA, USA
e-mail: vernescu@wpi.edu

1

Modeling of magnetorheological fluids has been, mostly, explored from a phenomenological point of view [3], by which a large class of admissible constitutive equations is derived, for which constitutive coefficients need to be either prescribed or experimentally obtained. These models also consider the Maxwell system decoupled from the fluid flow system. The theory of periodic homogenization, specifically designed to treat problems of highly heterogeneous and microstructure materials, was first used to derive effective models for magnetorheological fluids in [8, 9]. Improved effective models appeared a decade later in [16, 21] building upon the works in [8, 9]. The works in [8, 9, 16, 21] were further generalized in [15] where a new effective model was derived with a stress tensor that contained contributions from both the magnetic and the fluid components that depend on four different effective coefficients that can be numerically obtained from a series of local problems on the periodicity cell. In addition, one can better understand how the geometry of the periodic cell can be used [15] to numerically obtain the added nonlinear effect chain-like structures [3] have in strengthening the magnetorheological effect.

Magnetizable particles show a wide range of unusual magnetic properties. Surface particle magnetization is different from bulk volume magnetization which strongly influences the magnetorheological effect. It has been reported experimentally for magnetite nanoparticles (Fe_3O_4) an enhancement in saturation magnetization for particles of critical size up to ~ 10 nm beyond which the magnetization reduces [20]. The latter is attributed to surface effects becoming predominant as surface-to-volume ratio increases. Similar effects were reported by [10, 11] regarding the effective conductivity of composites with imperfect interfaces where the authors identified a critical radius R_{cr} for spherical particles such that for a polydisperse suspension of spheres when the mean radius lies below R_{cr} the effective conductivity of the composite lies below the conductivity of the enclosing matrix. In this work, we are able to numerically capture a similar effect where the effective magnetic coefficient value decreases as the surface-to-volume ratio increases for particles with 15% volume fraction.

The chapter is organized as follows: in Sect. 2, we introduce the hybrid model for magnetorheological suspensions as well as its effective counterpart together with the local problems. Section 3 is devoted to the numerical approximation of the local problems using viscosity and penalization methods. Section 4 computes the effective magnetic coefficients for different geometrical realizations ranging from a single particle chain to a cluster of particle chains. Lastly, Sect. 5 contains conclusions and some remarks.

1.1 Notation

Throughout the chapter, we are going to be using the following notation: I indicates the $n \times n$ identity matrix, $\mathbb{I}_C$ indicates the characteristic function over some set C, namely

$$\mathbb{I}_C(s) = \begin{cases} 0 & \text{if } s \in C \\ +\infty & \text{otherwise,} \end{cases}$$

bold symbols indicate vectors in two or three dimensions, regular symbols indicate tensors, $e(\boldsymbol{u})$ indicates the strain rate tensor defined by $e(\boldsymbol{u}) = \dfrac{1}{2}\left(\nabla \boldsymbol{u} + \nabla \boldsymbol{u}^\top\right)$, where often times we will use subscript to indicate the variable of differentiation. The inner product between matrices is denoted by $A{:}B = tr(A^\top B) = \sum_{ij} A_{ij} B_{ji}$, and throughout the chapter, we employ the Einstein summation notation for repeated indices.

2 Modeling Magnetorheological Suspensions

2.1 Hybrid Modeling of Magnetorheological Suspensions

We will next outline the coupled suspension model used in the homogenization process; it consists of the Newtonian fluid with rigid particles coupled flow system, coupled with Maxwell's equations.

As in the periodic homogenization framework, we first define the geometry of the suspension. We define $\Omega \subset \mathbb{R}^d$, $d \in \{2, 3\}$, to be a bounded open set with sufficiently smooth boundary $\partial\Omega$, $Y = [-1/2, 1/2)^d$ is the unit cube in $\mathbb{R}^d$, and $\mathbb{Z}^d$ is the set of all d-dimensional vectors with integer components. For every positive ϵ, let $N(\epsilon)$ be the set of all points $\ell \in \mathbb{Z}^d$ such that $\epsilon(\ell + Y)$ is strictly included in Ω, and denote by $|N(\epsilon)|$ their total number. Let T be the closure of an open connected set with sufficiently smooth boundary, compactly included in Y. For every $\epsilon > 0$ and $\ell \in N(\epsilon)$, we consider the set $T_\ell^\epsilon \subset\subset \epsilon(\ell + Y)$, where $T_\ell^\epsilon = \epsilon(\ell + T)$. The set T_ℓ^ϵ represents one of the rigid particles suspended in the fluid, and $S_\ell^\epsilon = \partial T_\ell^\epsilon$ denotes its surface (see Fig. 1). We now define the following subsets of Ω:

$$\Omega_{1\epsilon} = \bigcup_{\ell \in N^\epsilon} T_\ell^\epsilon, \quad \Omega_{2\epsilon} = \Omega \backslash \overline{\Omega}_{1\epsilon}.$$

In what follows, T_ℓ^ϵ will represent the magnetizable rigid particles, $\Omega_{1\epsilon}$ is the domain occupied by the rigid particles and $\Omega_{2\epsilon}$ the domain occupied by the surrounding fluid of viscosity $\nu \equiv 1$. We denote by $\partial\Omega$ the exterior boundary of Ω and $\partial\Omega_\epsilon := \cup_\ell S_\ell^\epsilon \cup \partial\Omega$.

By $\boldsymbol{n}$, we indicate the unit normal on the particle surface pointing outward, and by $[\![\cdot]\!]$, we indicate the jump discontinuity between the fluid and the rigid part.

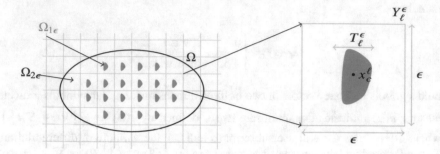

Fig. 1 Schematic of the periodic suspension of rigid magnetizable particles in non-magnetizable fluid

The magnetorheological problem considered in [15] after non-dimensionalizing and assuming that the flow is at low Reynolds numbers is the following:

$$-\text{div}\left(\sigma^\epsilon\right) = \mathbf{0}, \quad \text{where } \sigma^\epsilon = 2\,e(v^\epsilon) - p^\epsilon I \qquad\qquad \text{in } \Omega_{2\epsilon}, \qquad (1a)$$

$$\text{div}\left(v^\epsilon\right) = 0, \quad \text{div}\left(B^\epsilon\right) = 0, \quad \text{curl}\left(H^\epsilon\right) = \mathbf{0} \qquad \text{in } \Omega_{2\epsilon}, \qquad (1b)$$

$$e(v^\epsilon) = 0, \quad \text{div}\left(B^\epsilon\right) = 0, \quad \text{curl}\left(H^\epsilon\right) = \text{R}_\text{m}\, v^\epsilon \times B^\epsilon \qquad \text{in } \Omega_{1\epsilon}, \qquad (1c)$$

where from (1c), the following compatibility conditions result:

$$\text{div}\left(\text{R}_\text{m}\, v^\epsilon \times B^\epsilon\right) = 0, \quad \left\langle\text{R}_\text{m}\, v^\epsilon \times B^\epsilon \cdot n \mid 1\right\rangle_{H^{1/2}(S_\ell^\epsilon), H^{1/2}(S_\ell^\epsilon)} = 0. \qquad (2)$$

The interface and exterior boundary conditions are

$$\llbracket v^\epsilon \rrbracket = \mathbf{0}, \quad \llbracket B^\epsilon \cdot n \rrbracket = 0, \quad \llbracket n \times H^\epsilon \rrbracket = \mathbf{0} \qquad\qquad \text{on } S_\ell^\epsilon,$$
$$v^\epsilon = \mathbf{0}, \quad H^\epsilon \cdot n = c \cdot n \qquad\qquad\qquad\qquad \text{on } \partial\Omega. \qquad (3)$$

Here, v^ϵ represents the fluid velocity field, p^ϵ the pressure, $e(v^\epsilon)$ the strain rate, n the exterior normal to the particles, H^ϵ the magnetic field, μ^ϵ is the magnetic permeability of the material with $\mu^\epsilon(x) = \mu_1$ if $x \in \Omega_{1\epsilon}$ and $\mu^\epsilon(x) = \mu_2$ if $x \in \Omega_{2\epsilon}$ with $0 < \mu_2 < \mu_1$, B^ϵ is the magnetic induction, $B^\epsilon = \mu^\epsilon H^\epsilon$, and c is an applied constant magnetic field on the exterior boundary of the domain Ω, x_c^ℓ is the center of mass of the rigid particle T_ℓ^ϵ, α is the Alfven number, and R_m is the magnetic Reynolds number.

When the MR fluid is submitted to a magnetic field, the rigid particles are subjected to a force that makes them behave like a dipole aligned in the direction of the magnetic field. This force can be written in the form, $F^\epsilon := -\frac{1}{2}|H^\epsilon|^2 \nabla \mu^\epsilon$, where $|\cdot|$ represents the standard Euclidean norm. The force can be written in terms of the Maxwell stress $\tau_{ij}^\epsilon = \mu^\epsilon H_i^\epsilon H_j^\epsilon - \frac{1}{2}\mu^\epsilon H_k^\epsilon H_k^\epsilon \delta_{ij}$ as $F^\epsilon = \text{div}\left(\tau^\epsilon\right) - B^\epsilon \times \text{curl}\left(H^\epsilon\right)$. Since the magnetic permeability is considered constant in each phase, it follows that the force is zero in each phase. Therefore, we deduce that

$$\text{div}\left(\tau^{\epsilon}\right) = \begin{cases} 0 & \text{if } \boldsymbol{x} \in \Omega_{2\epsilon} \\ \boldsymbol{B}^{\epsilon} \times \text{curl}\left(\boldsymbol{H}^{\epsilon}\right) & \text{if } \boldsymbol{x} \in \Omega_{1\epsilon}. \end{cases} \tag{4}$$

Lastly, we remark that unlike the viscous stress σ^{ϵ}, the Maxwell stress is present in the entire domain Ω. Hence, we can write the balance of forces and torques for each particle as

$$0 = \int_{S_{\ell}^{\epsilon}} \sigma^{\epsilon} \boldsymbol{n} \, ds + \alpha \int_{S_{\ell}^{\epsilon}} [\![\tau^{\epsilon} \boldsymbol{n}]\!] \, ds - \alpha \int_{T_{\ell}^{\epsilon}} \boldsymbol{B}^{\epsilon} \times \text{curl}\left(\boldsymbol{H}^{\epsilon}\right) dx,$$

$$0 = \int_{S_{\ell}^{\epsilon}} \sigma^{\epsilon} \boldsymbol{n} \times (\boldsymbol{x} - \boldsymbol{x}_c^{\ell}) \, ds + \alpha \int_{S_{\ell}^{\epsilon}} [\![\tau^{\epsilon} \boldsymbol{n}]\!] \times (\boldsymbol{x} - \boldsymbol{x}_c^{\ell}) \, ds \tag{5}$$

$$- \alpha \int_{T_{\ell}^{\epsilon}} (\boldsymbol{B}^{\epsilon} \times \text{curl}\left(\boldsymbol{H}^{\epsilon}\right)) \times (\boldsymbol{x} - \boldsymbol{x}_c^{\ell}) \, dx.$$

The existence of a weak solution to the above model (1)–(5) relies [14] on the Altman-Shinbrot fixed point theorem [18] and certain a priori estimates for the velocity field $\boldsymbol{u}^{\epsilon}$ and the magnetic field $\boldsymbol{H}^{\epsilon}$, for sufficiently small R_m.

2.2 Effective Balance Equations

The effective balance equations, for the above model for an MR fluid, were obtained in [15] as the limit when $\epsilon \to 0$ of the system (1)–(5) under the assumption of quasi-neutrality [6] using two-scale expansions and have the following form:

$$\text{div}\left(\sigma^{hom} + \tau^{hom}\right) = \boldsymbol{0} \text{ in } \Omega,$$

$$\sigma^{hom} + \tau^{hom} = \left(-\bar{p}^0 + \frac{1}{d}(\beta_b - \beta_s)\left|\tilde{\boldsymbol{H}}^0\right|^2\right) I + \nu_s \, e(\boldsymbol{v}^0) + \beta_s \, \tilde{\boldsymbol{H}}^0 \otimes \tilde{\boldsymbol{H}}^0,$$

$$\text{div}\,\boldsymbol{v}^0 = 0 \text{ in } \Omega,$$

$$\text{div}\left(\mu^{hom}\,\tilde{\boldsymbol{H}}^0\right) = 0 \text{ in } \Omega,$$

$$\text{curl}\,\tilde{\boldsymbol{H}}^0 = R_m\,\boldsymbol{v}^0 \times \mu^{hom,s}\,\tilde{\boldsymbol{H}}^0 \text{ in } \Omega,$$

$$\tag{6}$$

with the boundary conditions:

$$\boldsymbol{v}^0 = \boldsymbol{0} \text{ on } \partial\Omega, \quad \tilde{\boldsymbol{H}}^0 \cdot \boldsymbol{n} = \boldsymbol{c} \cdot \boldsymbol{n} \text{ on } \partial\Omega. \tag{7}$$

We remark that a necessary compatibility condition that results from (6) is that

$$\text{div}(\text{R}_m \, \boldsymbol{v}^0 \times \mu^{hom,s} \, \widetilde{\boldsymbol{H}}^0) = 0 \text{ in } \Omega.$$ (8)

The effective stress tensor derived contains contribution from both the fluid and the magnetic field components, consisting an effective viscosity ν_s and four homogenized magnetic permeabilities, β_s, β_b, μ^{hom}, and $\mu^{hom,s}$ computed as the angular averaging of the tensors, $\nu_{ijm\ell}^{hom}$, $\beta_{ijm\ell}^{hom}$, μ_{ij}^{hom}, and $\mu_{ij}^{hom,s}$, which all depend on the geometry of the suspension, the volume fraction, the magnetic permeability μ, the Alfven number α, and the particles' distribution.

The effective coefficients $\nu_{ijm\ell}^{hom}$, $\beta_{ijm\ell}^{hom}$, μ_{ik}^{hom}, and $\mu_{ik}^{hom,s}$ are given by the formulas:

$$\nu_{ijm\ell}^{hom} = \int_{Y_f} 2\,e(\boldsymbol{B}^{ml} + \boldsymbol{\chi}^{ml}) : e(\boldsymbol{B}^{ij} + \boldsymbol{\chi}^{ij})\,d\boldsymbol{y},$$ (9)

$$\beta_{ijm\ell}^{hom} = \int_{Y_f} 2e(\boldsymbol{\xi}^{ml}) : e(\boldsymbol{B}^{ij} + \boldsymbol{\chi}^{ij})d\boldsymbol{y} + \alpha \int_{Y_f} \mu A^{ml} : e(\boldsymbol{B}^{ij} + \boldsymbol{\chi}^{ij})d\boldsymbol{y}$$

$$+ \alpha \int_Y \mu A_{ij}^{ml} d\boldsymbol{y},$$ (10)

$$\mu_{ik}^{hom} = \int_Y \mu \left(-\frac{\partial \phi^k}{\partial y_i} + \delta_{ik}\right) d\boldsymbol{y},$$ (11)

and

$$\mu_{ik}^{hom,s} = \int_T \mu \left(-\frac{\partial \phi^k}{\partial y_i} + \delta_{ik}\right) d\boldsymbol{y}.$$ (12)

Here, the effective coefficients are defined in terms of $\boldsymbol{\chi}^{m\ell}$, $\boldsymbol{\xi}^{ml}$, and ϕ^ℓ the solutions to the local problems formulated in the subsection below. We have also denoted $B_k^{ij} = \frac{1}{2}(y_i\,\delta_{jk} + y_j\,\delta_{ik}) - \frac{1}{d}y_k\,\delta_{ij}$ and the fourth-order tensor

$$A_{ij}^{ml} = \frac{1}{2} \left(A_{i\ell}\,A_{jm} + A_{j\ell}\,A_{im} - A_{mk}\,A_{\ell k}\,\delta_{ij}\right),$$ (13)

where $A_{i\ell}(\boldsymbol{y}) = \left(-\frac{\partial \phi^\ell(\boldsymbol{y})}{\partial y_i} + \delta_{i\ell}\right).$

2.3 Local Problems

In this subsection, we define the local problems to which the functions $\chi^{m\ell}$, ϕ^{ℓ}, and $\xi^{m\ell}$ are solutions to.

First, $\chi^{m\ell}$ that enters in the definition of the homogenized viscosity $v_{ij m\ell}^{hom}$ and of the magnetic permeability $\beta_{ij m\ell}^{hom}$ is a solution to the following local problem:

$$-\frac{\partial}{\partial y_j}\varepsilon_{ij}^{m\ell} = 0 \quad \text{in } Y_f,$$

$$\varepsilon_{ij}^{m\ell} = -p^{m\ell}\delta_{ij} + 2\left(C_{ij m\ell} + e_{ijy}(\chi^{m\ell})\right)$$

$$-\frac{\partial \chi_i^{m\ell}}{\partial y_i} = 0 \quad \text{in } Y_f, \tag{14}$$

$$\left[\!\left[\chi^{m\ell}\right]\!\right] = 0 \quad \text{on } S,$$

$$C_{ij m\ell} + e_{ijy}(\chi^{m\ell}) = 0 \quad \text{in } T,$$

$$\chi^{m\ell} \text{ is } Y - \text{periodic}, \quad \widetilde{\chi^{m\ell}} = \mathbf{0},$$

together with the balance of forces and torques,

$$\int_S \varepsilon_{ij}^{m\ell} n_j \, ds = 0, \quad \int_S \epsilon_{ijk}\, y_j\, \varepsilon_{kp}^{m\ell} n_p \, ds = 0, \tag{15}$$

where $C_{ij m\ell} = \frac{1}{2}(\delta_{im}\delta_{j\ell} + \delta_{i\ell}\delta_{jm}) - \frac{1}{d}\delta_{ij}\,\delta_{m\ell}$. We remark that if we define $B_k^{ij} = \frac{1}{2}(y_i\,\delta_{jk} + y_j\,\delta_{ik}) - \frac{1}{d}y_k\,\delta_{ij}$, then it immediately follows that $e_{ij}(\boldsymbol{B}^{m\ell}) = C_{ij m\ell}$.

The variational formulation of (14)–(15) is as follows: find $\chi^{m\ell} \in \mathcal{U}$ such that

$$\int_{Y_f} 2\, e_{ij}(\chi^{m\ell})\, e_{ij}(\boldsymbol{\phi} - \chi^{m\ell})\, d\boldsymbol{y} = 0, \quad \text{for all } \boldsymbol{\phi} \in \mathcal{U}, \tag{16}$$

where $\mathcal{U}$ is the closed, convex, non-empty subset of $H_{per}^1(Y)^d$ defined by

$$\mathcal{U} = \Big\{\boldsymbol{u} \in H_{per}^1(Y)^d \mid \text{div}\,(\boldsymbol{u}) = 0 \text{ in } Y_f, e_{ij}(\boldsymbol{u}) = -C_{ij m\ell} \text{ in } T,$$

$$\left[\!\left[\boldsymbol{u}\right]\!\right] = \mathbf{0} \text{ on } S, \widetilde{\boldsymbol{u}} = \mathbf{0} \text{ in } Y\Big\}. \tag{17}$$

The existence and uniqueness of a solution follows from classical theory of variational inequalities.

Second, the local problem defining ϕ^ℓ, which enters in the definition of all the effective magnetic permeabilities $\beta_{ijm\ell}^{hom}$, $\mu_{ik}^{hom,s}$, and μ_{ik}^{hom}, is

$$-\frac{\partial}{\partial y_i}\left(\mu\left(-\frac{\partial\phi^\ell}{\partial y_i}+\delta_{i\ell}\right)\right)=0 \quad \text{in } Y,$$

$$\left[\!\left[\mu\left(-\frac{\partial\phi^\ell}{\partial y_i}+\delta_{i\ell}\right)n_i\right]\!\right]=0 \quad \text{on } S, \tag{18}$$

$$\phi^\ell \text{ is } Y-\text{periodic,} \quad \widetilde{\phi^\ell}=0,$$

or in variational form: find $\phi^k \in \mathcal{W}=\left\{w\in H_{per}^1(Y)\mid \widetilde{w}=0\right\}$ such that

$$\int_Y \mu\frac{\partial\phi^k}{\partial y_i}\frac{\partial v}{\partial y_i}\,dy=\int_Y \mu\frac{\partial v}{\partial y_k}\,dy \text{ for any } v\in\mathcal{W}. \tag{19}$$

And the third local problem defining $\boldsymbol{\xi}^{m\ell}$, which enters in the definition of the homogenized permeability $\beta_{ijm\ell}^{hom}$, is

$$-\frac{\partial}{\partial y_j}\Sigma_{ij}^{m\ell}=0 \quad \text{in } Y_f,$$

$$\Sigma_{ij}^{m\ell}=-\pi^{m\ell}\delta_{ij}+2\,e_{ijy}(\boldsymbol{\xi}^{m\ell})$$

$$-\frac{\partial\xi_i^{m\ell}}{\partial y_i}=0 \quad \text{in } Y_f,$$

$$\left[\!\left[\boldsymbol{\xi}^{m\ell}\right]\!\right]=0 \quad \text{on } S, \tag{20}$$

$$e_{ijy}(\boldsymbol{\xi}^{m\ell})=0 \quad \text{in } T,$$

$$\boldsymbol{\xi}^{m\ell} \text{ is } Y-\text{periodic,} \quad \widetilde{\boldsymbol{\xi}^{m\ell}}=0,$$

with balance of forces and torques,

$$\int_S \Sigma_{ij}^{m\ell}\,n_j\,ds=0, \quad \int_S \epsilon_{ijk}\,y_j\left(\Sigma_{kp}^{m\ell}+\alpha\left[\!\left[\mu A_{kp}^{m\ell}\right]\!\right]\right)n_p\,ds=0. \tag{21}$$

We can formulate (20)–(21) variationally as follows: find $\boldsymbol{\xi}^{m\ell}\in\mathcal{V}$ such that

$$\int_{Y_f} 2\,e_{ijy}(\boldsymbol{\xi}^{m\ell})\,e_{ijy}(\boldsymbol{\phi})\,dy+\int_Y A_{ij}^{m\ell}\,e_{ijy}(\boldsymbol{\phi})\,dy=0, \text{ for all } \boldsymbol{\phi}\in\mathcal{V}, \tag{22}$$

where $\mathcal{V} = \left\{ v \in H^1_{\text{per}}(Y)^d \mid \text{div}\,(v) = 0 \text{ in } Y_f, \; e_y(u) = 0 \text{ in } T, \; [\![v]\!] = \mathbf{0} \text{ on } S, \; \tilde{v} = \mathbf{0} \text{ in } Y \right\}$ is a closed subspace of $H^1_{\text{per}}(Y)^d$. The existence and uniqueness follows from an application of the Lax–Milgram lemma. These equations indicate the contribution of the magnetic field, and the solution $\xi^{m\ell}$ depends, through the balance of forces and torques, on the solution of the local problem (19) and the effective magnetic permeability of the composite. We have defined the fourth-order tensor

$$A^{m\ell}_{ij} = \frac{1}{2} \left(A_{i\ell} A_{jm} + A_{j\ell} A_{im} - A_{mk} A_{\ell k} \delta_{ij} \right), \tag{23}$$

where $A_{i\ell}(y) = \left(-\frac{\partial \phi^\ell(y)}{\partial y_i} + \delta_{i\ell} \right)$. We remark that the only driving force that makes the solution $\xi^{m\ell}$ nontrivial in (22) is the rotation induced by the magnetic field through the fourth-order tensor $A^{m\ell}_{ij}$.

3 Computation of the Local Solutions: Penalization and Viscosity Methods

The goal of this section is to carry out calculations, using the finite element method, of the effective magnetic coefficients $\beta^{hom}_{ijk\ell}$ that characterize the magnetorheological effect in the presence of different geometrical realizations of chain particles.

In order to compute $\beta^{hom}_{ijk\ell}$, we must compute the local solutions $\chi^{m\ell}$, $\xi^{m\ell}$, and ϕ^k of the local problems (16), (22), and (19), respectively.

For all local problems, $\chi^{m\ell}$, $\xi^{m\ell}$, and ϕ^ℓ, we briefly describe how to implement the penalization method to enforce the rigid body motion of the particle and how to implement the viscosity method to enforce a zero average over the unit cell Y so that the local solutions can be uniquely determined.

We will discuss here how to implement the methods for the local solution $\chi^{m\ell}$ of (16) with the approximations of the other local solutions being similar. The solution to (16) can be classified as a minimum of the energy functional $\mathcal{J}$ in the following way: find $\chi^{m\ell} \in \mathcal{U}$ such that

$$\mathcal{J}(\chi^{m\ell}) = \min_{w \in H^1_{\text{per}}(Y)^d} \mathcal{J}(w), \tag{24}$$

where $\mathcal{J}(w) = \int_{Y_f} |e(w)|^2 \, dy + \mathbb{I}_{\mathcal{U}}(w)$. We approximate the solution to (24) by the following sequence of vector fields: find $\chi^{m\ell}_\lambda \in \overline{\mathcal{U}}$ such that

$$\mathcal{J}^\lambda(\chi^{m\ell}_\lambda) = \min_{u \in H^1_{\text{per}}(Y)^d} \mathcal{J}^\lambda(u), \tag{25}$$

where

$$
\mathcal{J}^\lambda(\boldsymbol{u}) = \int_{Y_f} |e(\boldsymbol{u})|^2 \, d\boldsymbol{y} + \frac{1}{2\lambda} \int_{Y_s} \left| e\left(\boldsymbol{u} + \boldsymbol{B}^{m\ell}\right) \right|^2 d\boldsymbol{y} + \mathbb{I}_{\overline{\mathcal{U}}}(\boldsymbol{u}), \tag{26}
$$

and

$$
\overline{\mathcal{U}} = \left\{ \boldsymbol{u} \in H_{\mathrm{per}}^1(Y)^d \mid \mathrm{div}\,(\boldsymbol{u}) = 0 \text{ in } Y_f,\ [\![\boldsymbol{u}]\!] = \boldsymbol{0} \text{ on } S,\ \tilde{\boldsymbol{u}} = \boldsymbol{0} \text{ in } Y \right\}. \tag{27}
$$

It is clear that $\mathcal{J}^\lambda$ is monotone in λ, and hence Γ converges in w-$H_{\mathrm{per}}^1(Y)^d$ to $\mathcal{J}$ as $\lambda \to 0$. To impose a zero average of the solution, we use the viscosity method to approximate $\boldsymbol{\chi}_\lambda^{m\ell}$ by $\boldsymbol{\chi}_{\lambda,\delta}^{m\ell}$ unique minimum of the following problem: find $\boldsymbol{\chi}_{\lambda,\delta}^{m\ell} \in \hat{\mathcal{U}}$ such that

$$
\mathcal{J}^{\lambda,\delta}(\boldsymbol{\chi}_{\lambda,\delta}^{m\ell}) = \min_{z \in H_{\mathrm{per}}^1(Y)^d} \mathcal{J}^{\lambda,\delta}(\boldsymbol{z}), \tag{28}
$$

where

$$
\mathcal{J}^{\lambda,\delta}(\boldsymbol{z}) = \int_{Y_f} |e(\boldsymbol{z})|^2 \, d\boldsymbol{y} + \frac{1}{2\lambda} \int_{Y_s} \left| e\left(\boldsymbol{z} + \boldsymbol{B}^{m\ell}\right) \right|^2 d\boldsymbol{y}
$$
$$
+ \frac{\delta}{2} \int_Y |\boldsymbol{z}|^2 \, d\boldsymbol{y} + \mathbb{I}_{\hat{\mathcal{U}}}(\boldsymbol{z}), \tag{29}
$$

and

$$
\hat{\mathcal{U}} = \left\{ \boldsymbol{u} \in H_{\mathrm{per}}^1(Y)^d \mid \mathrm{div}\,(\boldsymbol{u}) = 0 \text{ in } Y_f,\ [\![\boldsymbol{u}]\!] = \boldsymbol{0} \text{ on } S \right\}. \tag{30}
$$

By computing the Gateaux derivative of $\mathcal{J}^{\lambda,\delta}$, we obtain the following weak formulation:

$$
\int_{Y_f} 2\, e(\boldsymbol{\chi}_{\lambda,\delta}^{m\ell}) : e(\boldsymbol{\phi}) \, d\boldsymbol{y} + \frac{1}{\lambda} \int_{Y_s} e\left(\boldsymbol{\chi}_{\lambda,\delta}^{m,\ell} + \boldsymbol{B}^{m\ell}\right) : e(\boldsymbol{\phi}) \, d\boldsymbol{y}
$$
$$
+ \delta \int_Y \boldsymbol{\chi}_{\lambda,\delta}^{m\ell} \cdot \boldsymbol{\phi} \, d\boldsymbol{y} = 0, \tag{31}
$$

for any test function $\boldsymbol{\phi} \in \hat{\mathcal{U}}$. Using any constant as a test function, we can recover that $\boldsymbol{\chi}_\lambda^{m\ell} = \boldsymbol{0}$. Again, due to the monotonicity of $\mathcal{J}^{\lambda,\delta}$ in δ, we have that $\mathcal{J}^{\lambda,\delta}$ Γ

converges in w-$H^1_{\text{per}}(Y)^d$ to $\mathcal{J}^\lambda$. Hence, we have

$$\limsup_{\lambda \to 0} \limsup_{\delta \to 0} \mathcal{J}^{\lambda,\delta}(\chi^{m\ell}_{\lambda,\delta}) \leq \mathcal{J}(\chi^{m\ell}) \tag{32}$$

Using a diagonalization argument [1], there exists a map $\delta \mapsto \lambda(\delta)$ such that $\lim_{\delta \to 0} \lambda(\delta) = 0$ and

$$\limsup_{\delta \to 0} \mathcal{J}^{\lambda(\delta),\delta}(\chi^{m\ell}_{\lambda(\delta),\delta}) \leq \limsup_{\lambda \to 0} \limsup_{\delta \to 0} \mathcal{J}^{\lambda,\delta}(\chi^{m\ell}_{\lambda,\delta}). \tag{33}$$

A similar argument exists for the lim inf expression, but the inequality is reversed. Hence, the theory of Γ-convergence guarantees that our approximate solutions converge to the desired solutions. For more details, one can consult the works in [1], [13, Appendix A].

Thus, using the penalization and viscosity methods, we can compute the local solution $\xi^{m\ell}$. In Figs. 2 and 3, we plot $\xi^{m\ell}$ in the case of circular iron particles of 15% volume fraction and $\alpha = 1$ for different geometrical **chain** realizations.

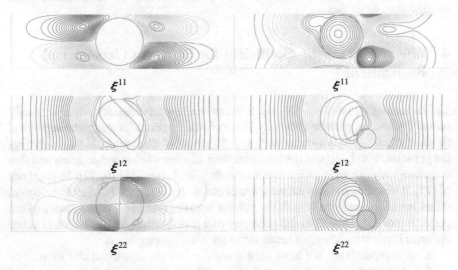

ξ^{11} $\qquad\qquad\qquad\qquad\qquad\qquad$ ξ^{11}

ξ^{12} $\qquad\qquad\qquad\qquad\qquad\qquad$ ξ^{12}

ξ^{22} $\qquad\qquad\qquad\qquad\qquad\qquad$ ξ^{22}

Fig. 2 Streamlines of the solution of the local problems for circular iron particles of 15% volume fraction and $\alpha = 1$ for different geometrical **chain** realizations. The left row showcases the streamlines for the local solution $\xi^{m\ell}$ in (22) for a single particle, while the row on the right showcases the streamlines for two particles

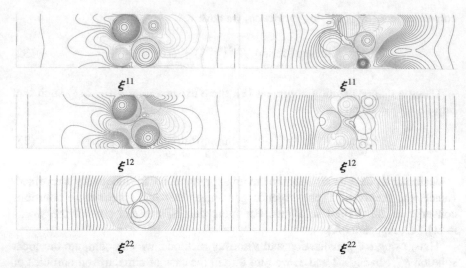

Fig. 3 Streamlines of the solution of the local problems for circular iron particles of 15% volume fraction and $\alpha = 1$ for different geometrical **chain** realizations. The left row showcases the streamlines for the local solution $\xi^{m\ell}$ in (22) for four particles, while the row on the right showcases the streamlines for six particles

4 Effective Magnetic Coefficient for Different Geometrical Realizations

Unlike regular suspensions for which the effective properties are dependent only on fluid viscosity, particle geometry, and volume fraction, for magnetorheological fluids of significance is also the particles' distribution. The magnetic field polarizes the particles, which align in the field direction, to form *chains* and *columns* and that contributes significantly to the increase in the yield stress [2, 19, 22]. In the work of [15], it was shown that in the presence of chain structures, the magnetorheological effect increases nonlinearly with the volume fraction (see Fig. 4). The choice of the periodic unit cell, as well as the geometry and distribution of particles, can lead to different *chain structures* and hence different effective properties.

In all computations, we have used a regular, symmetric, triangular mesh. For the $2 \times \frac{1}{2}$ periodic unit cell, we used 100×20 $P1$ elements for the rectangle and 100 $P1$ elements for the circular particle. Other geometries where the particle is an ellipse, for instance, present an interesting case on their own. Ellipses have a priori a preferred direction, i.e., they are anisotropic, and as a result, the effective coefficients will be anisotropic. Hence, when one discusses *chain structures* of ellipses, the *angle orientation* must be taken into account. However, we will not discuss such cases in the present work.

We remark that in the two-dimensional setting, the tensors entries $C_{ijmm} = 0$ and $\boldsymbol{B}^{mm} = \boldsymbol{0}$. As a consequence, of the linearity of the local problem (14), we have

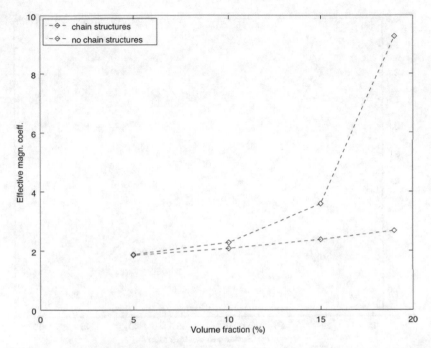

Fig. 4 Effective magnetic coefficient β_s plotted against volume fractions of 5%, 10%, 15%, and 19% for circular iron particles. The **red** color curve showcases the increase in the effective magnetic coefficient β_s under uniform particle distribution, while the **blue** color curve showcases the increase in the effective magnetic coefficient β_s in the presence of **chain structures**

$\chi^{mm} = 0$. Hence, $\nu_{mmii} = 0$, which implies that $\nu_b = 0$. Using a similar argument, we can similarly show that $\beta_{mmii} = 0$, which implies that $\beta_b = 0$.

The relative magnetic permeability of the iron particle (99.95% pure) was fixed throughout to be 2×10^5, while that of the fluid was set to 1 [4]. All the calculations were carried out using the software FreeFem++ [7].

For particles with a fixed volume fraction of 15%, we compute the effective magnetic coefficient β_s for different geometrical chain realizations. By breaking a single large particle into smaller particles, we achieve two objectives: (1) we obtain different chain structures formed from clustered particles and (2) we increase the surface area of the magnetic material while keeping the volume fraction the same. We considered four different geometrical realizations of chain structures with a single particle, two particles, four particles, and six particles as in Figs. 2 and 3.

Figure 5 showcases the effective coefficient β_s introduced in (6) versus the surface-to-volume ratio for different chain structures introduced in Figs. 2 and 3. We can readily observe that as the surface-to-volume ratio increases, the effective magnetic coefficient decreases, leading to a weaker magnetorheological effect for the overall suspension.

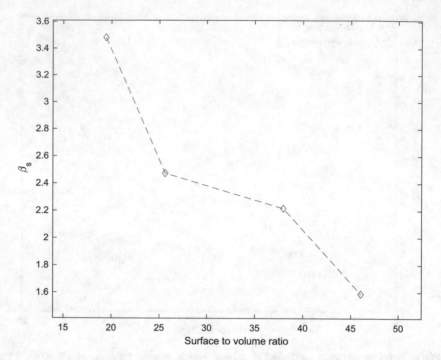

Fig. 5 Effective magnetic coefficient β_s plotted against surface-to-volume ratio for different realizations of chain particles with a constant volume fraction of 15%

5 Conclusions

Using the homogenization method, we have obtained, starting from a hybrid model for a suspension of rigid magnetizable particles in a Newtonian fluid, a model (6)–(7). The model has several novel features. First, it is a coupled system, and not partially uncoupled as in most of the literature; so, one cannot solve the Maxwell equations separately from the flow. Second, the pressure term in the constitutive equation in (6) exhibits a magnetic component, which is to our knowledge a novel in the magnetorheological literature. Third, unlike in phenomenological approaches, the coefficients of the various terms in the constitutive equation are defined precisely and depend on the material properties, geometry, and applied fields and thus can be computed numerically for different materials and structures. Fourth, it is to be remarked that the Maxwell equations in (6) have different magnetic permeabilities, which is again a novel part.

The model we obtained allows us to simulate the chain formation, which is the essence of the magnetorheological effect: the fact that the particles organize in chain structures is responsible for the non-Newtonian behavior of these materials, once a magnetic field is applied. In this chapter, we emphasize two phenomena. We show the nonlinearity of the effective magnetic coefficient β_s in the presence of chains vs.

uniform distribution as the particle volume fraction increases, and this translates in the increase of the apparent yield stress [15]. We also show the importance of total particle surface area as the value of β_s decreases by increasing the surface area and keeping the volume fraction constant.

Acknowledgments GN gratefully acknowledges the funding by the Deutsche Forschungsgemeinschaft (DFG, German Research Foundation) under Germany's Excellence Strategy—The Berlin Mathematics Research Center MATH+ (EXC-2046/1, project ID: 390685689) in project AA2-1. GN would also like to thank Konstantinos Danas of Laboratoire de Mécanique des Solides, CNRS, Ecole Polytechnique for the fruitful discussions. Additionally, the authors express their gratitude to the anonymous referees for their comments, suggestions, and corrections.

References

1. Attouch, H.: Variational Convergence for Functions and Operators. Pitman Advance Publishing Program, Boston (1984)
2. Bossis, G., Lacis, S., Meunier, A., Volkova, O.: Magnetorheological fluids. J. Magn. Magn. Mater. **252**, 224–228 (2002)
3. Brigadnov, I.A., Dorfmann, A.: Mathematical modelling of magnetorheological fluids. Continuum Mech. Thermodyn. **17**(1), 29–42 (2005)
4. Condon, E.H., Odishaw, H.: Handbook of Physics. McGraw-Hill, New York (1958)
5. de Vincente, J., Klingenberg, D.J., Hidalgo-Alvarez, R.: Magnetorheological fluids: a review. Soft Matter **7**, 3701–3710 (2011)
6. Duvaut, G., Lions, J.-L.: Les inéquations en mécanique et en physique. Travaux Recherches Math., Dunod, Paris (1972)
7. Hecht, F.: New development in freefem++. J. Numer. Math. **20**, 251–265 (2012)
8. Levy, T.: Suspension de particules solides soumises á des couples. J. Méch. Théor. App. Numéro spécial, pp. 53–71 (1985)
9. Levy, T., Hsieh, R.K.T.: Homogenization mechanics of a non-dilute suspension of magnetic particles. Int. J. Eng. Sci. **26**, 1087–1097 (1988)
10. Lipton, R., Vernescu, B.: Variational methods, size effects and extremal microgeometries for elastic composites with imperfect interface. Math. Mod. Meth. Appl. Sci. **5**(8), 1139–1173 (1995)
11. Lipton, R., Vernescu, B.: Composites with imperfect interface. Proc. R. Soc. Lond. A **452**, 329–358 (1996)
12. Liu, J., Flores, G.A., Sheng, R.: In-vitro investigation of blood embolization in cancer treatment using magnetorheological fluids. J. Magn. Magn. Mater. **225**(1–2), 209–217 (2001)
13. Nika, G.: Multiscale analysis of emulsions and suspensions with surface effects. Ph.D. thesis, Worcester Polytechnic Institute (2016)
14. Nika, G., Vernescu, B.: An existence result for a class of nonlinear magnetorheological composites (submitted)
15. Nika, G., Vernescu, B.: Multiscale modeling of magnetorheological suspensions. Z. Angew. Math. Phys. **71**, 1–19 (2020)
16. Perlak, J., Vernescu, B.: Constitutive equations for electrorheological fluids. Rev. Roumaine Math. Pures Appl. **45**, 287–297 (2000)
17. Rabinow, J.: The magnetic fluid clutch. AIEE Trans. **67**(17–18), 1308 (1948)
18. Shinbrot, M.: A fixed point theorem, and some applications. Arch. Rational Mech. Anal. **17**, 255–271 (1964)

19. Tao, R.: Super-strong magnetorheological fluids. J. Phys.: Condens. Matter **13**, 979–999 (2001)
20. Thapa, D., Palkar, V.R., Kurup, M.B., Malik, S.K.: Properties of magnetite nanoparticles synthesized through a novel chemical route. Mater. Lett. **58**, 2692–2694 (2004)
21. Vernescu, B.: Multiscale analysis of electrorheological fluids. Int. J. Modern Phys. B **16**, 2643–2648 (2002)
22. Winslow, W.M.: Induced fibration of suspensions. J. Appl. Phys. **20**, 1137–1140 (1949)

Multiscale Analysis of Myelinated Axons

Carlos Jerez-Hanckes, Isabel A. Martínez, Irina Pettersson,
and Volodymyr Rybalko

Abstract We consider a three-dimensional model for a myelinated neuron, which includes Hodgkin–Huxley ordinary differential equations to represent membrane dynamics at Ranvier nodes (unmyelinated areas). Assuming a periodic microstructure with alternating myelinated and unmyelinated parts, we use homogenization methods to derive a one-dimensional nonlinear cable equation describing the potential propagation along the neuron. Since the resistivity of intracellular and extracellular domains is much smaller than the myelin resistivity, we assume this last one to be a perfect insulator and impose homogeneous Neumann boundary conditions on the myelin boundary. In contrast to the case when the conductivity of the myelin is nonzero, no additional terms appear in the one-dimensional limit equation, and the model geometry affects the limit solution implicitly through an auxiliary cell problem used to compute the effective coefficient. We present numerical examples revealing the forecasted dependence of the effective coefficient on the size of the Ranvier node.

C. Jerez-Hanckes
Universidad Adolfo Ibáñez, Santiago, Chile
e-mail: carlos.jerez@uai.cl

I. A. Martínez
Pontificia Universidad Católica de Chile, Santiago, Chile
e-mail: iamartinez@uc.cl

I. Pettersson (✉)
University of Gävle, Gävle, Sweden
e-mail: irina.pettersson@hig.se

V. Rybalko
Institute for Low Temperature Physics and Engineering, Kharkiv, Ukraine
e-mail: vrybalko@nas.gov.ua

© The Author(s), under exclusive license to Springer Nature Switzerland AG 2021
P. Donato, M. Luna-Laynez (eds.), *Emerging Problems in the Homogenization of Partial Differential Equations*, SEMA SIMAI Springer Series 10,
https://doi.org/10.1007/978-3-030-62030-1_2

1 Introduction

A nerve impulse is the movement of the so-called action potential along a nerve fiber
in response to a stimulus such as touch, pain, heat, or cold. It is the way nerve cells
communicates with another cell so as to generate an adequate response. In their
work [5], Hodgkin and Huxley gave a plausible explanation of the physiological
process behind the excitability of nerve fibers and provided a phenomenological
mathematical model describing electric currents across axon membranes in terms of
ions' fluxes. The model describes how ionic currents' nonlinear dynamic behavior
depends on the potential difference across neurons' membranes and the so-called
gating variables, i.e., the probability for different ionic channels to be open or closed.
The jump of the potential across the membrane of each individual axon can be
modeled in the framework of the Hodgkin–Huxley (HH) model, but the alternating
myelinated and unmyelinated parts of the membrane present an obvious problem for
those attempting to describe its macroscopic response to the electrical stimulation.
In order to model and simulate the response of biological tissues to electrical
stimulation, one needs to know how signals propagate along single neurons and,
as the next step, how they influence each other in a bundle of axons.

Signal propagation along a neuron is portrayed by a cable equation usually
derived by modeling axons as cylinders comprised of segments with capacitances
and resistances combined in parallel [2, 5, 9, 13, 14]. The coefficients in such equation depend on the resistances and capacitances of Ranvier nodes and myelinated
parts, as well as some geometric parameters of the neuron such as the diameter
and nodal and internodal lengths. In [2, 10], the authors apply a formal two-scale
expansion to a one-dimensional model in order to show that a myelinated neuron
can be approximated by a homogeneous cable equation. However, these works do
not consider the derivation of the one-dimensional equation nor they provide any
justification of the formal approximations.

We derive a nonlinear cable equation for signal propagation along a myelinated
axon under the classical assumption that the conductivity of the myelin is zero,
i.e., the myelin is a perfect insulator. This assumption is justified by the fact that
the resistivity of the myelin is much larger than the resistivity of intracellular and
extracellular domains. This assumption does not lead to the appearance of a potential
in the limit equation as in our previous work [6]. Consequently, in this classical
case, we can suppress the geometrical assumptions on the myelin sheath such as
radial symmetry assumption and specific features at the points where myelin meets
intracellular domain. Our proof is in some sense simpler than the one in [6], but
since the intracellular domain is not a straight cylinder any more, an additional cell
problem appears (13). When the intracellular domain is a straight cylinder—the first
component of the normal vector is zero—the cell problem (13) has a trivial constant
solution and the effective coefficient coincides with one in [6].

The paper is organized as follows. In Sect. 2, we formulate our model problem
and present the main result in Theorem 1, with the rest of the paper devoted to
its proof. Section 3 presents a priori estimates for the potential u_ε and its jump

across Ranvier nodes, to then finally derive the one-dimensional effective problem in Sect. 4. Numerical solutions of the auxiliary cell problem and effective coefficient a^{eff} (see (11)) are provided in Sect. 5. We also present the computational results showing the dependence of the effective coefficient on the length of the Ranvier node and on the angle at which the myelin is attached to the intracellular domain.

2 Problem Setup

Let us consider a myelinated axon sparsely suspended in an extracellular medium. We assume that the axon has a periodic structure, containing myelinated and unmyelinated parts (nodes of Ranvier) as illustrated in Fig. 3.

2.1 Geometry

Given a bounded Lipschitz domain $\omega \subset \mathbb{R}^2$, we denote by Y (see Fig. 3) a periodicity cell:

$$Y = \left\{ y = (y_1, y') \in \mathbb{R}^3 : \quad y_1 \in \mathbf{T}^1, \ y' \in \phi(y_1)\omega \right\}.$$

Here, $\mathbf{T}^1$ is the one-dimensional torus and $\phi \subset C(\mathbf{T}^1)$. Let also ω_0 be a compact subset of ω in $\mathbb{R}^2$. The intracellular medium is defined as

$$Y_i = \left\{ y = (y_1, y') \in \mathbb{R}^3 : \quad y_1 \in \mathbf{T}^1, \ y' \in \phi_0(y_1)\omega_0 \right\},$$

where $\phi_0 \in C(\mathbf{T}^1)$. We assume that the cell Y is decomposed into two disjoint nonempty subdomains: an intracellular part Y_i and an extracellular medium Y_e as shown in the left-hand side of Fig. 3. The myelin sheath Y_m—depicted as white areas—is supposed to be perfectly insulating and modeled as voids. The extracellular part of the periodicity cell $Y_e = Y \setminus (Y_i \cup Y_m)$. The functions ϕ_0 and ϕ are supposed to be such that the boundary of Y_i does not touch the boundary of Y.

In case when $\phi(y_1)$ and $\phi_0(y_1)$ are constant, both the intracellular medium Y_i and the periodicity cell have constant cross-sections, while the factors $\phi(y_1)$ and $\phi_0(y_1)$ allow the cross-sections to vary.

We denote by Γ_m the boundary of the myelin sheath, and by Γ the Ranvier node—the unmyelinated part of the interface between Y_i and Y_e. The lateral boundary of Y is denoted by Σ. We assume that the boundary of the myelin part Γ_m is Lipschitz continuous.

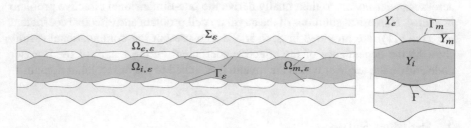

Fig. 1 Simplified geometry of the cross-section of a myelinated axon and the periodicity cell Y

We rescale the periodicity cell by a small parameter $\varepsilon > 0$ and translate it along the x_1-axis to form a thin periodic cylinder of thickness of order ε suspended in the thin extracellular medium with alternating myelinated and unmyelinated parts on the lateral boundary (cf. Fig. 1).

For simplicity, let us denote by $L \in \mathbb{N}$ an integer number of periods. The whole domain

$$\Omega_\varepsilon = \left\{(x_1, x') : \ x_1 \in (0, L), \ x' \in \varepsilon\phi\left(\frac{x_1}{\varepsilon}\right)\omega, \ \varepsilon > 0\right\}$$

is the union of the disjoint extracellular, intracellular domains, and Ranvier nodes: $\Omega_\varepsilon = \Omega_{i,\varepsilon} \cup \Omega_{e,\varepsilon} \cup \Gamma_\varepsilon$, wherein

$$\Omega_{i,\varepsilon} = \left\{(x_1, x') : \ x_1 \in (0, L), \ x' \in \varepsilon\phi_0\left(\frac{x_1}{\varepsilon}\right)\omega_0\right\}.$$

The lateral part of Ω_ε is denoted by Σ_ε. Let Γ_ε and $\Gamma_{m,\varepsilon}$ denote the Ranvier nodes and the boundary of the myelin, respectively. Note that we plot a cross-section of the domain Ω_ε. In $\mathbb{R}^3$, $\Omega_{e,\varepsilon}$ is connected, while Γ_ε and $\Omega_{m,\varepsilon}$ consist of a finite number of connected components.

Since the resistivity of intracellular and extracellular domains (e.g., 5.47×10^{-2} kOhm·cm) is much smaller than the resistivity of the myelin sheath (e.g., 7.4×10^5 kOhm·cm), a classical simplification is to assume that myelin is a perfect insulator. This implies that the areas Y_m constitute voids in our domain Y.

2.2 Governing Equations

Let u_ε^i and u_ε^e denote the electric potential in the intracellular and extracellular domains, respectively. We assume that the electric potential satisfies homogeneous Neumann boundary conditions on the lateral boundary Σ_ε and homogeneous Dirichlet ones at the bases $\Gamma_0^\varepsilon = \{x \in \Omega_\varepsilon : \ x_1 = 0\}$ and $\Gamma_L^\varepsilon = \{x \in \Omega_\varepsilon : \ x_1 = L\}$. Since we assume that the myelin acts as a perfect insulator, we impose homogeneous Neumann boundary conditions on its boundary.

The transmembrane potential is the potential jump across the axon membrane and will be denoted by $[u_\varepsilon] = u_\varepsilon^i - u_\varepsilon^e$, and u_ε denotes the potential $u_\varepsilon = u_\varepsilon^l$ in $\Omega_{i,\varepsilon}, l = i, e$. We assume conductivity to be a piecewise constant function $\sigma_\varepsilon = \sigma_l$ in $\Omega_{l,\varepsilon}, l = e, i$.

On Ranvier nodes, we assume continuity of currents (2) and HH dynamics for the transmembrane potential (3). Thus, the current through the membrane is a sum of the capacitive current $c_m \partial_t [u_\varepsilon]$, where c_m is the membrane capacitance per unit area, and the ionic current $I_{ion}([u_\varepsilon], g_\varepsilon)$ through the ion channels. The vector function g_ε is a vector of the so-called gating variables describing the probability of each particular ionic channel to be open or closed. Due to this, the gating variables have nonnegative components $0 \leq (g_\varepsilon)_j \leq 1$. The vector of gating variables satisfies an ordinary differential equation (ODE) $\partial_t g_\varepsilon = HH([u_\varepsilon], g_\varepsilon)$. In the classical HH model [5] there are three types of channels: a sodium (Na), a potassium (K), and a leakage channel, and, consequently, three gating variables (see Sect. 5 for explicit expressions). We assume that

(H1)   The function $I_{ion}(v, g)$ is linear w.r.t v and has the following form:

$$I_{ion}(v, g) = \sum_{j=1}^{m} H_j(g_j)(v - v_{r,j}),$$

where $g_{,j}$ is the jth component of g, $v_{r,j}$ is the jth component of the resting potential v_r, and H_j is positive, bounded, and Lipschitz continuous, that is,

$$|H_j(g_1) - H_j(g_2)| \leq L_1 |g_1 - g_2|.$$

The constant v_r is the reference constant voltage, and g_ε is a gate variable vector with nonnegative components $0 \leq (g_\varepsilon)_j \leq 1$, $j = \overline{1, m}$.

(H2)   The vector function $HH(g, v) = F(v) - \alpha g$, where F is a vector function with positive Lipschitz continuous components, and α is a diagonal $m \times m$ matrix with positive Lipschitz continuous entries.

(H3) $G_0 \in C(0, L)^m$ with components taking values between zero and one (as the corresponding g_ε).

We assume the homogeneous Dirichlet boundary condition for u_ε^e and for u_ε^i on the bases of the domain, i.e., when $x_1 = 0$ and $x_1 = L$. On the lateral boundary of Ω_ε, we assume the homogeneous Neumann boundary condition, with ν the unit normal exterior to $\Omega_{e,\varepsilon}$ on Σ and exterior to $\Omega_{i,\varepsilon}$ on Γ_ε.

The dynamics of the electric potential and the gating variables is then described by the following system of equations:

$$-\text{div}(\sigma_\varepsilon \Delta u_\varepsilon) = 0, \qquad\qquad\qquad (t, x) \in (0, T) \times \Omega_\varepsilon \setminus \Gamma_\varepsilon, \quad (1)$$

$$\sigma_e \nabla u_\varepsilon^e \cdot \nu = \sigma_i \nabla u_\varepsilon^i \cdot \nu, \qquad\qquad (t, x) \in (0, T) \times \Gamma_\varepsilon, \quad (2)$$

$$\varepsilon(c_m \partial_t [u_\varepsilon] + I_{ion}([u_\varepsilon], g_\varepsilon)) = -\sigma_i \nabla u_\varepsilon^i \cdot \nu, \quad (t, x) \in (0, T) \times \Gamma_\varepsilon, \quad (3)$$

$$\partial_t g_\varepsilon = HH([u_\varepsilon], g_\varepsilon), \qquad\qquad (t,x) \in (0,T) \times \Gamma_\varepsilon, \qquad (4)$$

$$[u_\varepsilon](x,0) = 0, \;\; g_\varepsilon(x,0) = G_0(x_1), \qquad\qquad x \in \Gamma_\varepsilon, \qquad (5)$$

$$\nabla u_\varepsilon^e \cdot v = 0, \qquad\qquad (t,x) \in (0,T) \times (\Gamma_{m,\varepsilon} \cup \Sigma_\varepsilon), \qquad (6)$$

$$u_\varepsilon = 0, \qquad\qquad (t,x) \in (0,T) \times (\Gamma_0^\varepsilon \cup \Gamma_L^\varepsilon). \qquad (7)$$

We will study the asymptotic behavior of u_ε, as $\varepsilon \to 0$, and derive a one-dimensional nonlinear cable equation describing the potential propagation along the axon.

To define a weak solution of (1)–(7), we will use test function $\phi \in L^\infty(0,T; H^1(\Omega_\varepsilon \setminus \Gamma_\varepsilon))$, $\partial_t[\phi] \in L^2(0,T; L^2(\Gamma_\varepsilon))$ such that $\phi = 0$ for $x_1 = 0$ and $x_1 = L$. The jump of ϕ across the Ranvier nodes is denoted by $[\phi]$, $[\phi] = (\phi^i - \phi^e)\big|_{\Gamma_\varepsilon}$. Then, the weak formulation corresponding to (1)–(7) reads as follows: find

$$u_\varepsilon \in L^\infty(0,T; H^1(\Omega_\varepsilon \setminus \Gamma_\varepsilon)), \quad \partial_t[u_\varepsilon] \in L^2(0,T; L^2(\Gamma_\varepsilon)),$$

satisfying $u_\varepsilon = 0$ for $x_1 = 0$ and $x_1 = L$ and the initial condition $[u_\varepsilon](0,x) = 0$, such that, for any test functions $\phi \in L^\infty(0,T; H^1(\Omega_\varepsilon \setminus \Gamma_\varepsilon))$, $\phi = 0$ for $x_1 = 0$ and $x_1 = L$, and for almost all $t \in (0,T)$, it holds

$$\varepsilon \int_{\Gamma_\varepsilon} c_m \partial_t[u_\varepsilon][\phi]\,ds + \int_{\Omega_\varepsilon \setminus \Gamma_\varepsilon} \sigma_\varepsilon \nabla u_\varepsilon \cdot \nabla \phi\,dx + \varepsilon \int_{\Gamma_\varepsilon} I_{ion}([u_\varepsilon], g_\varepsilon)[\phi]\,ds = 0. \qquad (8)$$

The vector of gating variables g_ε solves the following ODE:

$$\partial_t g_\varepsilon = HH([u_\varepsilon], g_\varepsilon), \;\; g_\varepsilon(0,x) = G_0(x_1).$$

Since HH is linear with respect to g_ε, we can solve the last ODE and obtain g_ε as a function—integral functional—of the jump $[u_\varepsilon]$:

$$\langle g_\varepsilon, v_\varepsilon \rangle = e^{-\int_0^t \alpha(v_\varepsilon(\zeta,x))d\zeta} \left(G_0(x) + \int_0^t F(v_\varepsilon(\tau,x)) e^{\int_0^\tau \alpha(v_\varepsilon(\zeta,x))d\zeta}\,d\tau \right).$$

Substituting this expression into (8) we obtain the weak formulation of (1)–(7) in terms of the potential u_ε and its jump $v_\varepsilon = [u_\varepsilon]$ across Γ_ε:

$$\varepsilon \int_{\Gamma_\varepsilon} c_m \partial_t v_\varepsilon[\phi]\,ds + \int_{\Omega_\varepsilon \setminus \Gamma_\varepsilon} \sigma_\varepsilon \nabla u_\varepsilon \cdot \nabla \phi\,dx + \varepsilon \int_{\Gamma_\varepsilon} I_{ion}(v_\varepsilon, \langle g_\varepsilon, v_\varepsilon \rangle)[\phi]\,ds = 0. \qquad (9)$$

2.3   Main Result

The main result of the paper is given by Theorem 1 describing the convergence of the transmembrane potential $[u_\varepsilon]$ and the gating variables g_ε to the unique solution of the following one-dimensional effective problem:

$$
\begin{aligned}
c_m \partial_t v_0 + I_{ion}(v_0, g_0) &= a^{\mathrm{eff}} \partial^2_{x_1 x_1} v_0, & (t, x_1) &\in (0, T) \times (0, L), \\
\partial_t g_0 &= HH(v_0, g_0), & (t, x_1) &\in (0, T) \times (0, L), & (10) \\
v_0(t, 0) &= v_0(t, L) = 0, & t &\in (0, T), \\
v_0(0, x_1) &= 0, \quad g_0(0, x_1) = G_0(x_1), & x_1 &\in (0, L).
\end{aligned}
$$

The effective coefficient a^{eff} is given by

$$
a^{\mathrm{eff}} = \frac{1}{|\Gamma|} \left(\left(\sigma_e \int_{Y_e} (\partial_{y_1} N_e + 1) \, dy \right)^{-1} + \left(\sigma_i \int_{Y_i} (\partial_{y_1} N_i + 1) \, dy \right)^{-1} \right)^{-1}, \quad (11)
$$

where the 1-periodic in y_1 functions, N_e and N_i solve the auxiliary cell problems:

$$
\begin{aligned}
-\Delta N_e(y) &= 0, & y &\in Y_e, \\
\nabla N_e \cdot \nu &= -\nu_1, & y &\in \Gamma \cup \Gamma_m \cup \Sigma, & (12) \\
N_e(y_1, y') & \text{ is periodic in } y_1;
\end{aligned}
$$

and

$$
\begin{aligned}
-\Delta N_i(y) &= 0, & y &\in Y_i, \\
\nabla N_i \cdot \nu &= -\nu_1, & y &\in \Gamma \cup \Gamma_m, & (13) \\
N_i(y_1, y') & \text{ is periodic in } y_1.
\end{aligned}
$$

Theorem 1 *The solutions $[u_\varepsilon]$ and g_ε of problems (1)–(7) converge to the solutions v_0 and g_0 of (10) in the following sense:*

$$
\sup_{t \in (0,T)} \varepsilon^{-1} \int_{\Gamma_\varepsilon} |[u_\varepsilon] - v_0|^2 ds \to 0, \quad \text{as } \varepsilon \to 0,
$$

$$
\sup_{t \in (0,T)} \varepsilon^{-1} \int_{\Gamma_\varepsilon} |g_\varepsilon - g_0|^2 ds \to 0, \quad \text{as } \varepsilon \to 0.
$$

To prove Theorem 1, we first derive a priori estimates in Sect. 3 (Lemma 2), then we prove the two-scale convergence of u_ε and its gradient (Lemma 3) and the convergence of $[u_\varepsilon]$ in appropriate spaces (Lemma 4). Finally, in Sect. 4 we pass to the limit in the weak formulation and derive the limit problem (10).

3   A Priori Estimates

The existence and uniqueness of a solution of (1)–(7) follows from the classical
semigroup theory (see, e.g., [11]). Its regularity is addressed in [3, 4, 7]. The proof
of Lemma 1 follows the lines of that in [6] (see Lemma 3.1).

Lemma 1 (Existence Result) *There exists a unique*

$$u_\varepsilon \in L^\infty(0, T; H^1(\Omega_\varepsilon \setminus \Gamma_\varepsilon)), \quad \partial_t v_\varepsilon = \partial_t[u_\varepsilon] \in L^2(0, T; L^2(\Gamma_\varepsilon))$$

such that $u_\varepsilon = 0$ for $x_1 = 0$ and $x_1 = L$, for any test functions $\phi \in$
$L^\infty(0, T; H^1(\Omega_\varepsilon \setminus \Gamma_\varepsilon))$, $\phi = 0$ for $x_1 = 0$ and $x_1 = L$, and for almost all $t \in (0, T)$

$$\varepsilon \int_{\Gamma_\varepsilon} c_m \partial_t v_\varepsilon[\phi] \, ds + \int_{\Omega_\varepsilon \setminus \Gamma_\varepsilon} \sigma_\varepsilon \nabla u_\varepsilon \cdot \nabla \phi \, dx + \varepsilon \int_{\Gamma_\varepsilon} I_{ion}(v_\varepsilon, \langle g_\varepsilon, v_\varepsilon \rangle)[\phi] \, ds = 0. \tag{14}$$

In order to pass to the limit in the weak formulation, we need to first obtain a
priori estimates that will guarantee the compactness of the solution in appropriate
spaces.

Lemma 2 (A Priori Estimates) *Let $(u_\varepsilon, g_\varepsilon)$ be a solution of (1)–(7). Denote $v_\varepsilon =$*
$[u_\varepsilon]$. Then, the following estimates hold:

(i) $\displaystyle \varepsilon^{-1} \int_{\Gamma_\varepsilon} |v_\varepsilon|^2 \, ds \le C, \quad t \in (0, T).$

(ii) $\displaystyle \varepsilon^{-1} \int_0^t \int_{\Gamma_\varepsilon} |\partial_\tau v_\varepsilon|^2 \, ds \, d\tau \le C, \quad t \in (0, T).$

(iii) $\displaystyle \varepsilon^{-2} \int_{\Omega_{i,\varepsilon} \cup \Omega_{e,\varepsilon}} (|u_\varepsilon|^2 + |\nabla u_\varepsilon|^2) \, dx \le C, \quad t \in (0, T).$

Proof We multiply (1) by u_ε and integrate by parts over $\Omega_\varepsilon \setminus \Gamma_\varepsilon$:

$$\frac{\varepsilon}{2} \frac{d}{dt} \int_{\Gamma_\varepsilon} c_m v_\varepsilon^2 \, ds + \varepsilon \int_{\Gamma_\varepsilon} I_{ion}(v_\varepsilon, \langle g_\varepsilon, v_\varepsilon \rangle) v_\varepsilon \, ds + \int_{\Omega_\varepsilon \setminus \Gamma_\varepsilon} \sigma_\varepsilon |\nabla u_\varepsilon|^2 \, dx = 0.$$

Integrating the last equality with respect to t and using (5) we get

$$\frac{\varepsilon}{2} \int_{\Gamma_\varepsilon} c_m v_\varepsilon^2 \, ds + \varepsilon \int_0^t \int_{\Gamma_\varepsilon} I_{ion}(v_\varepsilon, \langle g_\varepsilon, v_\varepsilon \rangle) v_\varepsilon \, ds \, d\tau + \int_0^t \int_{\Omega_\varepsilon \setminus \Gamma_\varepsilon} \sigma_\varepsilon |\nabla u_\varepsilon|^2 \, dx \, d\tau = 0. \tag{15}$$

Dividing the resulting identity by ε^2 (the scaling factor of the order $|\Omega_\varepsilon|$) and recalling the positivity of H we get

$$\frac{\varepsilon^{-1}}{2} \int_{\Gamma_\varepsilon} c_m v_\varepsilon^2 ds + \varepsilon^{-1} \int_0^t \int_{\Gamma_\varepsilon} \sum_j H(\langle g_\varepsilon, v_\varepsilon \rangle_j)(v_\varepsilon - v_{r,j})v_\varepsilon \, ds \, d\tau \le 0,$$

$$\frac{\varepsilon^{-1}}{2} \int_{\Gamma_\varepsilon} c_m v_\varepsilon^2 ds \le \frac{\varepsilon^{-1}}{2} \int_0^t \int_{\Gamma_\varepsilon} \sum_j H(\langle g_\varepsilon, v_\varepsilon \rangle_j)(v_{r,j})^2 \, ds \, d\tau \le C.$$

In this way estimate (i) is proved. Next, we derive an integral estimate for ∇u_ε from (15) and (i):

$$\varepsilon^{-2} \int_0^t \int_{\Omega_\varepsilon \backslash \Gamma_\varepsilon} \sigma_\varepsilon |\nabla u_\varepsilon|^2 \, dx dt \le C.$$

Let us now multiply (1) by $\partial_t u_\varepsilon$ and integrate by parts over $\Omega_\varepsilon \backslash \Gamma_\varepsilon$:

$$\varepsilon^{-1} \int_{\Gamma_\varepsilon} c_m |\partial_t v_\varepsilon|^2 ds + \varepsilon^{-1} \int_{\Gamma_\varepsilon} I_{ion}(v_\varepsilon, \langle g_\varepsilon, v_\varepsilon \rangle) \partial_t v_\varepsilon \, ds$$

$$+ \frac{\varepsilon^{-2}}{2} \frac{d}{dt} \int_{\Omega_\varepsilon \backslash \Gamma_\varepsilon} \sigma_\varepsilon |\nabla u_\varepsilon|^2 \, dx = 0.$$

Integrating with respect to t gives

$$\varepsilon^{-1} \int_0^t \int_{\Gamma_\varepsilon} c_m |\partial_\tau v_\varepsilon|^2 ds d\tau + \varepsilon^{-1} \int_0^t \int_{\Gamma_\varepsilon} I_{ion}(v_\varepsilon, \langle g_\varepsilon, v_\varepsilon \rangle) \partial_\tau v_\varepsilon \, ds d\tau$$

$$+ \frac{\varepsilon^{-2}}{2} \int_{\Omega_\varepsilon} \sigma_\varepsilon |\nabla u_\varepsilon|^2 \, dx = \frac{\varepsilon^{-2}}{2} \int_{\Omega_\varepsilon \backslash \Gamma_\varepsilon} \sigma_\varepsilon |\nabla u_\varepsilon|^2 \Big|_{t=0} dx. \tag{16}$$

Since v_ε is a strict solution, we can choose $\phi = u_\varepsilon$ and set $t = 0$ in (8). Then, $\nabla u_\varepsilon |_{t=0} = 0$. By (H1), the boundedness of H, and the estimate (i), we derive

$$\varepsilon^{-1} \int_0^t \int_{\Gamma_\varepsilon} c_m |\partial_\tau v_\varepsilon|^2 ds d\tau + \varepsilon^{-1} \int_0^t \int_{\Gamma_\varepsilon} \sum_j H(\langle g_\varepsilon, v_\varepsilon \rangle_j)(v_\varepsilon - v_{r,j}) \partial_\tau v_\varepsilon \, ds d\tau \le 0,$$

$$\varepsilon^{-1} \int_0^t \int_{\Gamma_\varepsilon} c_m |\partial_\tau v_\varepsilon|^2 ds d\tau \le C \varepsilon^{-1} \int_0^t \int_{\Gamma_\varepsilon} (v_\varepsilon - v_{r,j}) \partial_\tau v_\varepsilon \, ds d\tau,$$

$$\varepsilon^{-1} \int_0^t \int_{\Gamma_\varepsilon} |\partial_\tau v_\varepsilon|^2 ds d\tau \le C \varepsilon^{-1} \int_0^t \int_{\Gamma_\varepsilon} (v_\varepsilon - v_{r,j})^2 \, ds d\tau \le C.$$

Estimate (ii) is proved. Estimates (16) and (ii) imply that

$$\varepsilon^{-2} \int_{\Omega_\varepsilon \backslash \Gamma_\varepsilon} \sigma_\varepsilon |\nabla u_\varepsilon|^2 \, dx \le C, \quad t \in (0, T). \tag{17}$$

Since u_ε satisfies the homogeneous Dirichlet boundary condition for $x_1 = 0$, the Friedrichs's inequality is valid for u_ε in Ω_ε^i and Ω_ε^e leading to (iii).

$\square$

When passing to the limit, as $\varepsilon \to 0$, we will use the notion of the two-scale convergence. Let us recall the definition.

Definition 1 We say that $u_\varepsilon(t, x)$ converges two-scale to $u_0(t, x_1, y)$ in $L^2(0, T; L^2(\Omega_{l,\varepsilon})), l = i, e$, if

(i) $\varepsilon^{-2} \int_0^T \int_{\Omega_{l,\varepsilon}} |u_\varepsilon|^2 dx\, dt < \infty$.

(ii) For any $\phi(t, x_1) \in C(0, T; L^2(0, L))$, $\psi(y) \in L^2(Y_l)$, we have

$$\lim_{\varepsilon \to 0} \varepsilon^{-2} \int_0^T \int_{\Omega_{l,\varepsilon}} u_\varepsilon(x) \phi(t, x_1) \psi\left(\frac{x}{\varepsilon}\right) dx\, dt$$

$$= \int_0^T \int_0^L \int_{Y_l} u_0(t, x_1, y) \phi(t, x_1) \psi(y)\, dy\, dx_1\, dt,$$

for some function $u_0 \in L^2(0, T; L^2((0, L) \times Y))$.

Definition 2 We say that $v_\varepsilon(t, x)$ converges two-scale to $v_0(t, x_1, y)$ in $L^2(0, T; L^2(\Gamma_\varepsilon))$ if it holds that

(i) $\varepsilon^{-1} \int_0^T \int_{\Gamma_\varepsilon} v_\varepsilon^2\, ds\, dt < \infty$.

(ii) For any $\phi(t, x_1) \in L^\infty(0, T; L^2(0, L))$, $\psi(y) \in L^2(\Gamma)$, we have

$$\lim_{\varepsilon \to 0} \varepsilon^{-1} \int_0^T \int_{\Gamma_\varepsilon} v_\varepsilon(x) \phi(t, x_1) \psi\left(\frac{x}{\varepsilon}\right) ds_x\, dt$$

$$= \int_0^T \int_0^L \int_\Gamma v_0(t, x_1, y) \phi(t, x_1) \psi(y)\, ds_y\, dx_1\, dt,$$

for some function $v_0 \in L^2\left(0, T; L^2((0, L) \times \Gamma)\right)$.

Lemma 3 *Let u_ε be a solution of (1)–(7). Denote by $\mathbf{I}_{\Omega_{l,\varepsilon}}$ the characteristic functions of $\Omega_{l,\varepsilon}, l = i, e$. Then, up to a subsequence,*

(i) $[u_\varepsilon]$ converges two-scale to $v_0(t, x_1, y)$ in $L^2(0, T; L^2(\Gamma_\varepsilon))$.

(ii) $\partial_t[u_\varepsilon]$ converges two-scale to $\partial_t v_0(t, x_1, y)$ in $L^2(0, T; L^2(\Gamma_\varepsilon))$.

(iii) $\mathbf{I}_{\Omega_{l,\varepsilon}} u_\varepsilon$ converges two-scale to $|Y_l| u_0^l(t, x_1)$ in $L^2(0, T; L^2(\Omega_{l,\varepsilon}))$.

(iv) $\mathbf{I}_{\Omega_{l,\varepsilon}} \nabla u_\varepsilon$ converges two-scale to $(\partial_{x_1} u_0^l(t, x_1) \mathbf{e_1} + \nabla_y w^l(t, x_1, y)$ in $(L^2(0, T; L^2(\Omega_{l,\varepsilon})))$. Here $\mathbf{e_1} = (1, 0, 0) \in \mathbf{R}^3$, $w^l \in L^2(0, T; L^2(0, L) \times H^1(Y))$.

For the proof, we refer to [1] for two-scale convergence on periodic surfaces (on Γ_ε) and to [15] and [12] for two-scale convergence in thin structures and dimension reduction.

One of the technical difficulties in the present paper is the passage to the limit in the integral over Γ_ε containing a nonlinear function since we need to ensure a strong convergence of v_ε in an appropriate sense. In the next lemma, we show that v_ε can be approximated by a piecewise constant function $\tilde{v}_\varepsilon(t, x_1)$, which in its turn converges, up to a subsequence, to a function $v_0(t, x_1) \in L^\infty(0, T; H^1(0, L))$ uniformly on $[0, T]$, as $\varepsilon \to 0$.

Lemma 4 *Let u_ε be a solution of (1)–(7). Then, there exists a function*

$$\tilde{v}_\varepsilon(t, x_1) \in L^\infty(0, T; H^1(0, L)) \cap H^1(0, T; L^2(0, L))$$

such that, it holds

(i) For $t \in (0, T)$, the function $\tilde{v}_\varepsilon$ approximates $[u_\varepsilon]$:

$$\int_{\Gamma_\varepsilon} |\tilde{v}_\varepsilon - [u_\varepsilon]|^2 ds \le C\varepsilon \int_{\Omega_{i,\varepsilon} \cup \Omega_{e,\varepsilon}} |\nabla u_\varepsilon|^2 dx.$$

(ii) There exists $v_0(t, x_1) \in L^\infty(0, T; L^2(0, L))$ such that along a subsequence $\tilde{v}_\varepsilon$ converges to $v_0(t, x_1)$ uniformly on $[0, T]$, as $\varepsilon \to 0$.

Proof Let us cover Ω_ε by a union of overlapping cells $\varepsilon \tilde{Y}_k$ as shown in Fig. 2 so that each cell contains two Ranvier nodes. The Ranvier node that belongs to the intersection $\varepsilon \tilde{Y}_k \cap \varepsilon \tilde{Y}_{k+1}$ is denoted by $\varepsilon \Gamma_k$. The intra- and extracellular parts of $\tilde{Y}_k$ are referred to as $\tilde{Y}_{i,k}$ and $\tilde{Y}_{e,k}$, respectively.

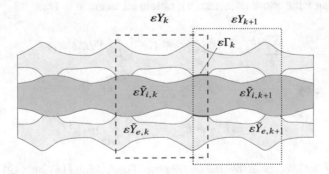

Fig. 2 Overlapping cells $\varepsilon \tilde{Y}_k$ and $\varepsilon \tilde{Y}_{k+1}$

Let us show that the difference between the mean values of $[u_\varepsilon]$ over $\varepsilon\Gamma_k$ and $\varepsilon\Gamma_{k+1}$ is small. Let

$$\bar{u}^l_{\varepsilon,k} = \frac{1}{|\varepsilon\Gamma|} \int_{\varepsilon\Gamma_k} u^l_\varepsilon ds, \quad l = i, e.$$

For each $\varepsilon\tilde{Y}_{l,k}, l = i, e$, due to the Poincaré inequality, we have

$$\int_{\varepsilon\tilde{Y}_{l,k}} |u^l_\varepsilon - \bar{u}^l_{\varepsilon,k}|^2 dx \le C\varepsilon^2 \int_{\varepsilon\tilde{Y}_{l,k}} |\nabla u^l_\varepsilon|^2 dx,$$

with C independent of ε. Considering traces on Γ_k, by a simple scaling argument one has

$$\int_{\varepsilon\Gamma_k} |u^l_\varepsilon - \bar{u}^l_{\varepsilon,k}|^2 ds \le C\varepsilon^{-1} \left(\int_{\varepsilon\tilde{Y}_{l,k}} |u^l_\varepsilon - \bar{u}^l_{\varepsilon,k}|^2 dx + \varepsilon^2 \int_{\varepsilon\tilde{Y}_{l,k}} |\nabla u^l_\varepsilon|^2 dx \right)$$

$$\le C\varepsilon \int_{\varepsilon\tilde{Y}_{l,k}} |\nabla u^l_\varepsilon|^2 dx, \quad l = i, e.$$

(18)

Then, the difference between $\bar{u}_{\varepsilon,k}$ and $\bar{u}_{\varepsilon,k+1}$ is estimated as follows:

$$\left| \bar{u}^l_{\varepsilon,k} - \bar{u}^l_{\varepsilon,k+1} \right|^2 \le \frac{2}{|\varepsilon\tilde{Y}_{l,k} \cap \varepsilon\tilde{Y}_{l,k+1}|} \int_{\varepsilon\tilde{Y}_{l,k} \cap \varepsilon\tilde{Y}_{l,k+1}} \left(|u^l_\varepsilon - \bar{u}^l_{\varepsilon,k}|^2 + |u^l_\varepsilon - \bar{u}^l_{\varepsilon,k+1}|^2 \right) dx$$

$$\le C\varepsilon^{-1} \int_{\varepsilon\tilde{Y}_{l,k} \cup \varepsilon\tilde{Y}_{l,k+1}} |\nabla u^l_\varepsilon|^2 dx.$$

Adding up in k the above estimates, we obtain an estimate in Ω^l_ε:

$$\sum_k |\bar{u}^l_{\varepsilon,k} - \bar{u}^l_{\varepsilon,k+1}|^2 \le C\varepsilon^{-1} \int_{\Omega^l_\varepsilon} |\nabla u^l_\varepsilon|^2 dx.$$

(19)

Let us denote by

$$\bar{v}_{\varepsilon,k} = \bar{u}^i_{\varepsilon,k} - \bar{u}^e_{\varepsilon,k} = \frac{1}{|\varepsilon\Gamma|} \int_{\varepsilon\Gamma_k} [u_\varepsilon] ds$$

the jump of the average across the membrane. Then, using (18) and (19) yields

$$\int_{\varepsilon\Gamma_k} \left| [u_\varepsilon] - \bar{v}_{\varepsilon,k} \right|^2 ds \le C\varepsilon \int_{\varepsilon\tilde{Y}_{i,k} \cup \varepsilon\tilde{Y}_{e,k}} |\nabla u_\varepsilon|^2 dx,$$

$$\sum_k |\bar{v}_{\varepsilon,k} - \bar{v}_{\varepsilon,k+1}|^2 \le C\varepsilon^{-1} \int_{\Omega^i_\varepsilon \cup \Omega^e_\varepsilon} |\nabla u_\varepsilon|^2 dx.$$

(20)

Bounds (20) show that $[u_\varepsilon]$ in each cell $\varepsilon \tilde{Y}_k$ is close to a constant $\bar{v}_{\varepsilon,k}$, and the difference between $\bar{v}_{\varepsilon,k}$ and $\bar{v}_{\varepsilon,k+1}$ is small due to (iii) in Lemma 2.

Next, we construct a piecewise linear function $\tilde{v}_\varepsilon(t, x_1)$ interpolating values $\bar{v}_{\varepsilon,k}$ linearly and show that

$$\int_0^L |\tilde{v}_\varepsilon|^2 dx_1 \leq C, \qquad t \in (0, T), \tag{21}$$

$$\int_0^L |\partial_{x_1} \tilde{v}_\varepsilon|^2 dx_1 \leq C, \quad t \in (0, T), \tag{22}$$

$$\int_0^T \int_0^L |\partial_t \tilde{v}_\varepsilon|^2 dx_1 dt \leq C. \tag{23}$$

Indeed, (21) and (22) follow directly from (20) and (i) and (ii) in Lemma 2:

$$\int_0^L |\tilde{v}_\varepsilon|^2 dx_1 = \sum_k \int_{-\varepsilon/2}^{\varepsilon/2} \Big| \frac{\bar{v}_{\varepsilon,k} + \bar{v}_{\varepsilon,k+1}}{2} + x_1 \frac{\bar{v}_{\varepsilon,k+1} - \bar{v}_{\varepsilon,k}}{\varepsilon} \Big|^2 dx_1$$

$$\leq C \sum_k \varepsilon (|\bar{v}_{\varepsilon,k}|^2 + |\bar{v}_{\varepsilon,k+1}|^2) \leq C \frac{1}{|\varepsilon \Gamma|} \int_{\Gamma_\varepsilon} [u_\varepsilon]^2 ds \leq C. \tag{24}$$

Estimate (22) is proved in a similar way using (20):

$$\int_0^L |\partial_{x_1} \tilde{v}_\varepsilon|^2 dx_1 \leq C \sum_k \int_{-\varepsilon/2}^{\varepsilon/2} \Big| \frac{\bar{v}_{\varepsilon,k} - \bar{v}_{\varepsilon,k+1}}{\varepsilon} \Big|^2 dx_1$$

$$\leq C \varepsilon^{-1} \sum_k |\bar{v}_{\varepsilon,k} - \bar{v}_{\varepsilon,k+1}|^2$$

$$\leq C \varepsilon^{-2} \int_{\Omega_\varepsilon^i \cup \Omega_\varepsilon^e} |\nabla u_\varepsilon|^2 dx \leq C.$$

Let us prove (23). Differentiating $\bar{v}_{\varepsilon,k}$ with respect to t, using the Cauchy–Schwarz inequality yields

$$|\partial_t \bar{v}_{\varepsilon,k}|^2 = \Big| \frac{1}{|\varepsilon \Gamma_k|} \int_{\varepsilon \Gamma_k} \partial_t [u_\varepsilon] ds \Big|^2 \leq \frac{1}{|\varepsilon \Gamma_k|} \int_{\varepsilon \Gamma_k} (\partial_t [u_\varepsilon])^2 ds.$$

Similar to (24), estimate (23) follows from the last bound and (ii) in Lemma 2. Estimate (i) in the current lemma follows from (20).

The uniform convergence on $(0, T)$ of the constructed piecewise linear approximation is given by the Arzelà–Ascoli theorem. Indeed, the precompactness is guaranteed for $\tilde{v}_\varepsilon$ by (21) and (22), while the equicontinuity property follows from (23):

$$\tilde{v}_\varepsilon(t + \Delta t) - \tilde{v}_\varepsilon(t) = \int_t^{t+\Delta t} \partial_\tau \tilde{v}_\varepsilon(\tau) d\tau,$$

$$\varepsilon^{-1} \int_0^L |\tilde{v}_\varepsilon(t + \Delta t) - \tilde{v}_\varepsilon(t)|^2 dx \le \int_0^L \left(\int_t^{t+\Delta t} \partial_\tau \tilde{v}_\varepsilon(\tau) d\tau \right)^2 dx_1$$

$$\le \Delta t \int_0^L \int_t^{t+\Delta t} |\partial_\tau \tilde{v}_\varepsilon(\tau)|^2 d\tau dx_1 \le \Delta t.$$

The proof is complete.   □

4 Justification of Macroscopic Model

Let us denote $v_\varepsilon = [u_\varepsilon]$. Using Lemmata 3 and 4, we will pass to the limit in the weak formulation of (1)–(7):

$$\varepsilon^{-1} \int_0^T \int_{\Gamma_\varepsilon} (c_m \partial_t v_\varepsilon + I_{ion}(v_\varepsilon, \langle g_\varepsilon, v_\varepsilon \rangle))[\phi] \, dx dt$$

$$+\varepsilon^{-2} \int_0^T \int_{\Omega_\varepsilon \setminus \Gamma_\varepsilon} \sigma_\varepsilon \nabla u_\varepsilon \cdot \nabla \phi \, dx dt = 0, \tag{25}$$

where $\phi(t, x) \in L^\infty(0, T; H^1(\Omega_\varepsilon \setminus \Gamma_\varepsilon))$ such that $\phi = 0$ for $x_1 = 0$ and $x_1 = L$.

For the functions $U_i(t, x_1), U_e(t, x_1) \in C(0, T; C_0^\infty(0, L))$ and $V_i, V_e(t, x_1, y) \in C(0, T; C_0^\infty(0, L) \times H^1(Y))$, we construct the following test function:

$$\phi_\varepsilon(t, x) = (U_i(t, x_1) + \varepsilon V_i(t, x_1, y))\chi_{\Omega_{i,\varepsilon}} + \chi_{\Omega_{e,\varepsilon}}(U_e(t, x_1) + \varepsilon V_e\left(t, x_1, \frac{x}{\varepsilon}\right)),$$

where $\chi_{\Omega_{l,\varepsilon}}$ is the characteristic function of $\Omega_{l,\varepsilon}, l = i, e$.

Note that the jump of ϕ_ε on Γ_ε converges strongly in $L^2(\Gamma_\varepsilon)$ to $U_i(t, x_1) - U_e(t, x_1)$. Substituting ϕ_ε into (25) we get

$$\varepsilon^{-1} \int_0^T \int_{\Gamma_\varepsilon} (c_m \partial_t v_\varepsilon + I_{ion}(v_\varepsilon, \langle g_\varepsilon, v_\varepsilon \rangle))[\phi_\varepsilon] \, ds dt \tag{26}$$

$$+\varepsilon^{-2} \int_0^T \int_{\Omega_{i,\varepsilon}} \sigma_i \nabla u_\varepsilon^i \cdot (\mathbf{e_1} \partial_{x_1} U_i + \varepsilon \nabla V_i\left(x_1, \frac{x}{\varepsilon}\right)) dx dt \tag{27}$$

$$+\varepsilon^{-2} \int_0^T \int_{\Omega_{e,\varepsilon}} \sigma_e \nabla u_\varepsilon^e \cdot (\mathbf{e_1} \partial_{x_1} U_e + \varepsilon \nabla V_e\left(x_1, \frac{x}{\varepsilon}\right)) dx dt \tag{28}$$

$$= I_{1\varepsilon} + I_{2\varepsilon} + I_{3\varepsilon} = 0.$$

We will pass to the limit, as $\varepsilon \to 0$, in each integral $I_{k\varepsilon}$, $k = 1, 2, 3$, given by (26)–(28).

Since $[\phi_\varepsilon]$ on Γ_ε converges strongly in $L^2(\Gamma_\varepsilon)$ to $U_i(t, x_1) - U_e(t, x_1)$ and $\partial_t v_\varepsilon$ converges two-scale (weakly) in $L^2(0, T; L^2(\Gamma_\varepsilon))$ and uniformly on $(0, T)$ to $v_0(t, x_1)$, we can pass to the limit in (26) and obtain

$$I_{1\varepsilon} = \varepsilon^{-1} \int_0^T \int_{\Gamma_\varepsilon} (c_m \partial_t v_\varepsilon + I_{ion}(v_\varepsilon, \langle g_\varepsilon, v_\varepsilon \rangle))[\phi_\varepsilon] \, ds \, dt$$

$$\xrightarrow[\varepsilon \to 0]{} |\Gamma| \int_0^T \int_0^L (c_m \partial_t v_0 + I_{ion}(v_0, \langle g_0, v_0 \rangle))(U_i - U_e) \, dx_1 dt.$$

To pass to the two-scale limit in (27)–(28), we use (iv) in Lemma 3 and get

$$I_{2\varepsilon} = \varepsilon^{-2} \int_0^T \int_{\Omega_{i,\varepsilon}} \sigma_i \nabla u_\varepsilon^i \cdot (\mathbf{e_1} \partial_{x_1} U_i + \nabla_y V_i(x_1, \frac{x}{\varepsilon}) + \varepsilon \partial_{x_1} V_i(x_1, \frac{x}{\varepsilon})) dx dt$$

$$\xrightarrow[\varepsilon \to 0]{} \int_0^T \int_0^L \int_{Y_i} \sigma_i (\mathbf{e_1} \partial_{x_1} u_0^i + \nabla_y w^i) \cdot (\mathbf{e_1} \partial_{x_1} V_i(t, x_1) + \nabla_y V_i(t, x_1, y)) \, dy \, dx_1 dt.$$

$$I_{3\varepsilon} = \varepsilon^{-2} \int_0^T \int_{\Omega_{e,\varepsilon}} \sigma_e \nabla u_\varepsilon^e \cdot (\mathbf{e_1} \partial_{x_1} U_e + \nabla_y V_e(x_1, \frac{x}{\varepsilon}) + \varepsilon \partial_{x_1} V_e(x_1, \frac{x}{\varepsilon})) dx dt$$

$$\xrightarrow[\varepsilon \to 0]{} \int_0^T \int_0^L \int_{Y_e} \sigma_e (\mathbf{e_1} \partial_{x_1} u_0^e + \nabla_y w^e) \cdot (\mathbf{e_1} \partial_{x_1} V_e(t, x_1) + \nabla_y V_e(t, x_1, y)) \, dy \, dx_1 dt.$$

Thus, we obtain a weak formulation of the effective problem:

$$|\Gamma| \int_0^T \int_0^L (c_m \partial_t v_0 + I_{ion}(v_0, \langle g_0, v_0 \rangle))(U_i - U_e) \, dx_1 dt$$

$$+ \int_0^T \int_0^L \int_{Y_i} \sigma_i (\mathbf{e_1} \partial_{x_1} u_0^i + \nabla_y w^i) \cdot (\mathbf{e_1} \partial_{x_1} U_i(t, x_1) + \nabla_y V_i(t, x_1, y)) \, dy \, dx_1 dt$$

$$+ \int_0^T \int_0^L \int_{Y_e} \sigma_e (\mathbf{e_1} \partial_{x_1} u_0^e + \nabla_y w^e) \cdot (\mathbf{e_1} \partial_{x_1} U_e(t, x_1) + \nabla_y V_e(t, x_1, y)) \, dy \, dx_1 dt = 0.$$

Consequently, computing the variation of the left-hand side of the last equality with respect to V_i, V_e, U_i, and U_e gives the representations $V_i(t, x_1, y) = N_i(y)\partial_{x_1} U_i(t, x_1)$, $V_e(t, x_1, y) = N_e(y)\partial_{x_1} U_e(t, x_1)$, the cell problems (12) and (13), and the two one-dimensional equations

$$|\Gamma|(c_m \partial_t v_0 + I_{ion}(v_0, \langle g_0, v_0 \rangle)) = \int_{Y_i} \sigma_e |\mathbf{e_1} + \nabla_y N_i|^2 \, \partial_{x_1 x_1}^2 u_0^i \, dy, \qquad (29)$$

$$|\Gamma|(c_m \partial_t v_0 + I_{ion}(v_0, \langle g_0, v_0 \rangle)) = -\int_{Y_e} \sigma_e |\mathbf{e_1} + \nabla_y N_e|^2 \, \partial_{x_1 x_1}^2 u_0^e \, dy. \qquad (30)$$

Introducing (11) and adding up (30) and (29) yield (10). The proof of Theorem 1 is complete.

5 Numerical Example

The goal of this numerical example is to see how the effective coefficient defined by (11) varies with respect to the area of Γ. We consider a rotationally symmetric geometry as illustrated in Fig. 1. Since the first component of the normal to Y_i is zero in this case, the problem reduces to solving the auxiliary cell problem (12) in the extracellular domain Y_e. For this, we use a finite element approximation. Having N_e, we compute the effective coefficient a^{eff}, whose formula in this cylindrical geometry becomes

$$a^{\text{eff}} = \frac{1}{|\Gamma|} \left(\left(\sigma_e \int_{Y_e} (\partial_{y_1} N_e + 1) \, dy \right)^{-1} + \left(\sigma_i |Y_i| \right)^{-1} \right)^{-1}, \tag{31}$$

as the cell problem (13) has a constant solution in this case and $\partial_{y_1} N_i = 0$. The effective coefficient has units $S \cdot cm^2$, that is the units of the conductivity S/cm multiplied by cm^3. The conductivities of the extra- and intracellular domains are assumed to be $\sigma_e = 20$ mS/cm and $\sigma_i = 5$ mS/cm. The node–node separation might vary between $500 \, \mu$m and $1500 \, \mu$m (see [8]), and we will take the period $L = 1250 \, \mu$m. To make the period equal to one, we need to rescale the domain by L. For example, the radius of the node is 1.8μm, so $r_0 = 1.8/1250$.

Table 1 contains the geometric parameters of the domain.

The values for the effective coefficient a^{eff} computed for angles $\alpha = \beta = \pi/2$ (angles of the myelin attachment) and for different values of the length of the

Table 1 Geometric parameters in μm

$R_0 L$	$(r_0 + m)L$	$r_0 L$	$l \cdot L$
9	5.75	1.8	1

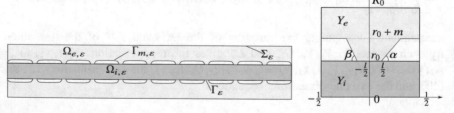

Fig. 3 The cross-section of half of the periodic cell, where $Y = \left(-\frac{1}{2}, \frac{1}{2} \right) \times D_{R_0}$ and $Y_i = \left(-\frac{1}{2}, \frac{1}{2} \right) \times D_{r_0}$, with D_{R_0} and D_{r_0} being the open disks in $\mathbb{R}^2$ of radius R_0 and r_0, respectively

Table 2 Results of the effective coefficient a^{eff} for different values of l, the length of the Ranvier node. l is in μm

$l \cdot L$	0.5	1	2	4	8	16
a^{eff}	1.1	$5.5 \cdot 10^{-1}$	$2.8 \cdot 10^{-1}$	$1.4 \cdot 10^{-1}$	$6.9 \cdot 10^{-2}$	$3.5 \cdot 10^{-2}$
$l \cdot L$	32	64	128	256	512	1024
a^{eff}	$1.7 \cdot 10^{-2}$	$8.6 \cdot 10^{-3}$	$4.3 \cdot 10^{-3}$	$2.2 \cdot 10^{-3}$	$1.1 \cdot 10^{-3}$	$5.4 \cdot 10^{-4}$

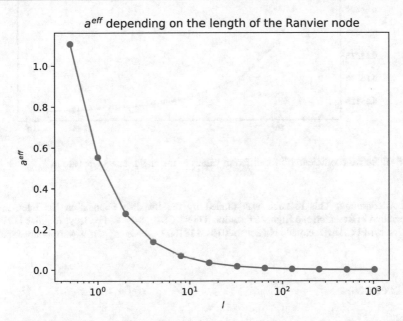

Fig. 4 Effective coefficient a^{eff} for different values of l, the length of the Ranvier node. l is in μm

Table 3 Results of the effective coefficient a^{eff} for different values of the angle. The angles are in degrees and the results presented with six significant digits

α°	0.4	0.5	1	2	5
a^{eff}	0.555167	0.554720	0.553897	0.553516	0.553297
α°	10	20	46	95	
a^{eff}	0.553223	0.553187	0.553165	0.553153	

Ranvier node $l \cdot L$ are shown in Table 2 and Fig. 4. It can be observed that a^{eff} decreases when l increases.

We also analyze how the effective coefficient depends on the angles of the myelin attachment. The results of the computations are presented in Table 3 and Fig. 5. One can see that the variation of a^{eff} is not significant, but the effective coefficient clearly decreases when the angles go to zero.

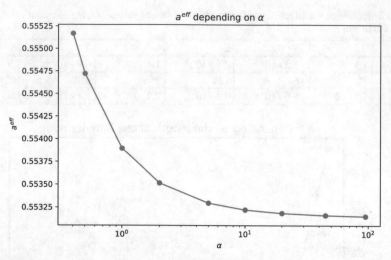

Fig. 5 Effective coefficient a^{eff} for different values of the angle. The angles are in °

Acknowledgments This research was funded by the Swedish Foundation for International Cooperation in Research and Higher Education STINT CS2018-7908, Fondecyt Regular 1171491, and Conicyt-PFCHA/Doctorado Nacional/2018-21181809, whose support is warmly appreciated.

References

1. Allaire, G., Damlamian, A., Hornung, U.: Two-scale convergence on periodic surfaces and applications. In: Mathematical Modelling of Flow through Porous Media (1995)
2. Basser, P.J.: Cable equation for a myelinated axon derived from its microstructure. Med. Biol. Eng. Comput. **31**(1), S87–S92 (1993)
3. Henríquez, F., Jerez-Hanckes, C.: Multiple traces formulation and semi-implicit scheme for modelling biological cells under electrical stimulation. ESAIM: M2AN **52**(2), 659–703 (2018)
4. Henríquez, F., Jerez-Hanckes, C., Altermatt, F.: Boundary integral formulation and semi-implicit scheme coupling for modeling cells under electrical stimulation. Numer. Math. **136**(1), 101–145 (2016)
5. Hodgkin, A.L., Huxley, A.F.: A quantitative description of membrane current and its application to conduction and excitation in nerve. J. Physiol. **117**(4), 500 (1952)
6. Jerez-Hanckes, C., Pettersson, I., Rybalko, V.: Derivation of cable equation by multiscale analysis for a model of myelinated axons. Discrete Contin. Dyn. Syst. Ser. B **25**(3), 815–839 (2020)
7. Matano, H., Mori, Y.: Global existence and uniqueness of a three-dimensional model of cellular electrophysiology. Discrete Contin. Dyn. Syst. **29**, 1573–1636 (2011)
8. McIntyre, C.C., Richardson, A.G., Grill, W.M.: Modeling the excitability of mammalian nerve fibers: influence of afterpotentials on the recovery cycle. J. Neurophysiol. **87**(2), 995–1006 (2002)
9. Meffin, H., Tahayori, B., Sergeev, E.N., Mareels, I.M.Y., Grayden, D.B., Burkitt, A.N.: Modelling extracellular electrical stimulation: III. derivation and interpretation of neural tissue equations. J. Neural Eng. **11**(6), 065004 (2014)

10. Meunier, C., d'Incamps, B.L.: Extending cable theory to heterogeneous dendrites. Neural Comput. **20**(7), 1732–1775 (2008)
11. Pazy, A.: Semigroups of Linear Operators and Applications to Partial Differential Equations, vol. 44. Springer Science & Business Media, New York (2012)
12. Pettersson, I.: Two-scale convergence in thin domains with locally periodic rapidly oscillating boundary. Differential Equations Appl. **9**(3), 393–412 (2017)
13. Rall, W.: Time constants and electrotonic length of membrane cylinders and neurons. Biophys. J. **9**(12), 1483–1508 (1969)
14. Rattay, F.: Electrical Nerve Stimulation. Theory, Experiments and Applications. Springer, Vienna (1990)
15. Zhikov, V.: On an extension and an application of the two-scale convergence method. Mat. Sb. **191**(7), 31–72 (2000)

Homogenization for Alternating Boundary Conditions with Large Reaction Terms Concentrated in Small Regions

María-Eugenia Pérez-Martínez

Abstract We consider a homogenization problem for the Laplace operator posed in a bounded domain of the upper half-space, a part of its boundary being in contact with the plane. On this part, the boundary conditions alternate from Neumann to Robin, being of Dirichlet type outside. The Robin conditions are imposed on small regions periodically placed along the plane and contain a *Robin parameter* that can be very large. The period tends to zero, and we provide all the possible homogenized problems, depending on the relations between the three parameters: period, size of the small regions, and Robin parameter. We address the convergence of the solutions in the most critical case where a *non-constant capacity coefficient* arises in the *strange term*.

1 Introduction

This work deals with the homogenization of alternating boundary conditions of Robin type with large parameters where, depending on the relations between the parameters of the problem, different strange terms may appear. From the geometrical viewpoint, it belongs to a large class of boundary homogenization problems studied for a long time in the literature of applied mathematics for different operators. We mention a few works related to scalar problem such as [1, 2, 4–8, 13, 18, 22, 23, 27], and [29]; some of them have introduced keywords such as critical sizes and critical relations between parameters (see [6, 22], and [29] for further references, in this connection).

We consider a homogenization problem for the Laplace operator posed in a domain Ω of the upper half-space $\mathbb{R}^{3+}$, a part of its boundary Σ being in contact with the plane $\{x_3 = 0\}$. A Dirichlet boundary condition is imposed out of Σ. On Σ,

M.-E. Pérez-Martínez (✉)
Departamento de Matemática Aplicada y Ciencias de la Computación, Universidad de Cantabria, Santander, Spain
e-mail: meperez@unican.es

© The Author(s), under exclusive license to Springer Nature Switzerland AG 2021
P. Donato, M. Luna-Laynez (eds.), *Emerging Problems in the Homogenization of Partial Differential Equations*, SEMA SIMAI Springer Series 10,
https://doi.org/10.1007/978-3-030-62030-1_3

37

the boundary conditions alternate from Neumann to Robin. The same operator and geometrical configuration here considered but with alternating boundary conditions of Neumann type (or Robin type) and Dirichlet type has been addressed, for instance, in [4] and [26]; cf. [3] and [19] for the elasticity system. In this chapter, the Robin boundary conditions have coefficients $\beta(\varepsilon)\alpha(x)$ that concentrate in small regions T^ε along Σ, while $\beta(\varepsilon)$ can be very large.

The model under consideration (6) may represent the scalar version of a *Winkler foundation* comprised of a block of an elastic material that has a part of its boundary $(\partial\Omega \setminus \Sigma)$ clamped to a rigid support, while the other part Σ lies partially on a series of "small springs" with elastic coefficients containing the parameter $\beta(\varepsilon)$. The Robin conditions recall the elastic response or the reaction of the media; cf. [14] and [15] in this connection. See [11–13], and [8] for an extensive bibliography on the homogenization for perforated domains along manifolds with large parameters on the boundary conditions. We also consider the associated spectral problem.

The small regions T^ε mentioned above have a diameter $O(r_\varepsilon)$ and are placed along the plane at a distance $O(\varepsilon)$ between them, see Figure 1. Here, ε and r_ε are two parameters that converge toward zero, $r_\varepsilon \ll \varepsilon$, while $\beta(\varepsilon)$ in fact can be large or small or of order 1, as $\varepsilon \to 0$. Three different relations between these parameters play an important role when describing the asymptotic behavior of the solution (cf. (1)–(3)). We find *critical relations* between parameters for which different *strange terms* arise in the homogenized Robin boundary condition. This homogenized condition is intermediate between Dirichlet and Neumann boundary conditions.

A critical relation between ε, r_ε, and $\beta(\varepsilon)$ appears provided by $\beta^* > 0$, where

$$\lim_{\varepsilon\to 0} \beta(\varepsilon) r_\varepsilon^2 \varepsilon^{-2} = \beta^*. \tag{1}$$

For large r_ε, cf. (2) with $r_0 = +\infty$, the homogenized Robin condition contains somewhat *averaged coefficients*, cf. (25), obtained from the coefficients of the

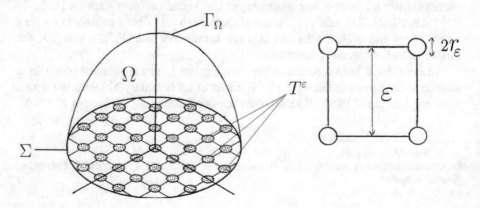

Fig. 1 Geometrical configuration of the problem

homogenization problem $\beta(\varepsilon)\alpha(x)\chi_{T^\varepsilon}(x)$, where χ_{T^ε} denotes the characteristic function of $\bigcup T^\varepsilon$. Notice that $\beta^* > 0$ is obtained when the total area of the regions T^ε, $O(\varepsilon^{-2}r_\varepsilon^2)$, multiplied by *the Robin parameter* $\beta(\varepsilon)$ is of order 1, in such a way that a critical size of T^ε corresponds to each Robin parameter $\beta(\varepsilon)$, namely $r_\varepsilon = O(\beta(\varepsilon)^{-1/2}\varepsilon)$, while a critical Robin parameter $\beta(\varepsilon) = O(\varepsilon^2 r_\varepsilon^{-2})$ corresponds to each size r_ε.

Other key relations between parameters are given by

$$\lim_{\varepsilon\to 0} r_\varepsilon \varepsilon^{-2} = r_0 \tag{2}$$

and

$$\lim_{\varepsilon\to 0} r_\varepsilon \beta(\varepsilon) = \beta^0. \tag{3}$$

In the case where $r_0 > 0$, we deal with the *classical critical size* of the regions T^ε, which was obtained in the case of Dirichlet conditions on T^ε instead of Robin (cf., e.g., [4] and [26]). The most critical situation happens when $r_0 > 0$ and $\beta^0 > 0$, which also amounts to $r_0 > 0$ and $\beta^* > 0$. In this case, cf. (19), the strange term contains a function $C^e(x)$, which is referred to as *extended capacity*, cf. (20), which depends on the Robin coefficient $\alpha(x)$ of the homogenization problem in a nontrivial way. To have an idea of this dependence, one may compare with the problem for perforated domains by balls, where we can compute explicitly this dependence (see, e.g., [8]).

Another relation that implies a strange term deals with $r_0 > 0$ and $\beta^0 = +\infty$ (also $\beta^* = +\infty$). This amounts, asymptotically, to Dirichlet conditions on the regions T^ε, and consequently the capacity term is constant, ignoring both $\beta(\varepsilon)$ and $\alpha(x)$, cf. (22): capacity and strange term coincide with those obtained, e.g., in the Appendix of [26].

Finally, the structure of the chapter is as follows. Section 2 contains the setting of the homogenization problem, the associated spectral problem, and a priori estimates for solutions and eigenvalues. Section 3 contains the list of homogenized problems with the corresponding local problems, depending on the values of r_0, β^0, and β^*. They are obtained using matched asymptotic expansions. Section 4 contains the setting of the local problems in unbounded domains and some estimates for solutions which are key points to show the convergence. Section 5 addresses the convergence in the most critical situation (cf. Remarks 1 and 2 in this connection). Finally, also for the most critical case, Sect. 6 addresses the spectral convergence.

2   The setting of the problem

Let Ω be a bounded domain of $\mathbb{R}^3$ situated in the upper half-space $\mathbb{R}^{3+} = \{x \in \mathbb{R}^3 : x_3 > 0\}$, with a Lipschitz boundary $\partial\Omega$. Let Σ be the part of the boundary in contact with the plane $\{x_3 = 0\}$, which is assumed to be non-empty, and let Γ_Ω be the rest of the boundary of Ω: $\partial\Omega = \overline{\Gamma}_\Omega \cup \overline{\Sigma}$. Let T denote a bounded domain of the plane $\{x_3 = 0\}$ with a smooth boundary. Without any restriction, we can assume that both Σ and T contain the origin of coordinates.

Let ε be a small parameter $\varepsilon \ll 1$. We consider r_ε an order function such that $r_\varepsilon \ll \varepsilon$. For $k = (k_1, k_2) \in \mathbb{Z}^2$, we denote by $\widetilde{x}_k^\varepsilon$ the point of the plane $\{x_3 = 0\}$ of coordinates $\widetilde{x}_k^\varepsilon = (k_1\varepsilon, k_2\varepsilon, 0)$ and by $T_{\widetilde{x}_k}^\varepsilon$ the homothetic domain of T of ratio r_ε after translation to the point $\widetilde{x}_k^\varepsilon$, namely, the set

$$T_{\widetilde{x}_k}^\varepsilon = \widetilde{x}_k^\varepsilon + r_\varepsilon T.$$

If there is no ambiguity, we shall write $\widetilde{x}_k$ instead of $\widetilde{x}_k^\varepsilon$ and T^ε instead of $T_{\widetilde{x}_k}^\varepsilon$.

In this way, for a fixed ε, we have constructed a grid of squares in the plane $\{x_3 = 0\}$ whose vertices are in the regions T^ε. Let the set $\mathcal{J}^\varepsilon$ denote $\mathcal{J}^\varepsilon = \{k \in \mathbb{Z}^2 : T_{\widetilde{x}_k}^\varepsilon \subset \Sigma\}$, while N_ε denotes the number of elements of $\mathcal{J}^\varepsilon$:

$$N_\varepsilon \approx \frac{|\Sigma|}{\varepsilon^2} = O(\varepsilon^{-2}). \tag{4}$$

Finally, if no confusion arises, we denote by $\bigcup T^\varepsilon$ the union of all the T^ε contained in Σ. Also, in what follows, $x = (x_1, x_2, x_3)$ denotes the usual Cartesian coordinates, while by $\hat{x} = (x_1, x_2)$ we refer to the two first components of $x \in \mathbb{R}^3$.

Let us consider a function $\alpha(x)$ of $C(\overline{\Omega})$ satisfying $\alpha(x) \geq \widehat{\alpha} > 0$ and uniformly Lipschitz in $\overline{\Omega}$, namely,

$$|\alpha(x) - \alpha(x')| \leq \widehat{c}|x - x'|, \quad \forall x, x' \in \overline{\Omega}, \tag{5}$$

with $\widehat{\alpha}$ and $\widehat{c}$ given constants. Let $f \in L^2(\Omega)$ and u^ε be the solution of the following homogenization problem:

$$\begin{cases} -\Delta u^\varepsilon = f & \text{in } \Omega, \\ u^\varepsilon = 0 & \text{on } \Gamma_\Omega, \\ \dfrac{\partial u^\varepsilon}{\partial n} = 0 & \text{on } \Sigma \setminus \bigcup T^\varepsilon, \\ \dfrac{\partial u^\varepsilon}{\partial n} + \beta(\varepsilon)\alpha(x)u^\varepsilon = 0 & \text{on } \bigcup T^\varepsilon, \end{cases} \tag{6}$$

where n stands for the unit outer normal to Ω along Σ, and $\beta(\varepsilon)$ is a positive parameter. It can range from very large to very small and is referred to as *the Robin parameter*.

The variational formulation of (6) reads: Find $u^\varepsilon \in \mathbf{V}$ satisfying

$$\int_\Omega \nabla u^\varepsilon . \nabla v \, dx + \beta(\varepsilon) \int_{\bigcup T^\varepsilon} \alpha \, u^\varepsilon v \, d\hat{x} = \int_\Omega f v \, dx, \quad \forall v \in \mathbf{V}, \tag{7}$$

where the space $\mathbf{V}$ is obtained by completion of $\{v \in C^1(\overline{\Omega}) : v = 0 \text{ on } \Gamma_\Omega\}$ with respect to the Dirichlet norm.

The left-hand side of (7) defines a bilinear, symmetric, continuous, and coercive form on $\mathbf{V} \subset \mathbf{L}^2(\Omega)$ and, therefore, there is a unique solution $u^\varepsilon \in \mathbf{V}$. On account of the Poincaré inequality and the positivity of the function α, u^ε satisfies

$$\int_\Omega |\nabla u^\varepsilon|^2 \, dx \leq C \quad \text{and} \quad \beta(\varepsilon) \int_{\bigcup T^\varepsilon} (u^\varepsilon)^2 \, d\hat{x} \leq C, \tag{8}$$

where C is a constant independent of ε. Hence, for any sequence, we can extract a subsequence, still denoted by ε such that

$$u^\varepsilon \to u^0 \text{ in } H^1(\Omega)\text{-}weak, \quad \text{as } \varepsilon \to 0, \tag{9}$$

for some u^0 that we will identify with the solution of a certain homogenized problem.

Depending on the different relations for the parameters ε, r_ε, and $\beta(\varepsilon)$, namely, on the values β^*, r_0, and β^0 in (1), (2), and (3), respectively, in Section 3, we state all the possible homogenized problems. For the sake of brevity, in Section 5, we only provide the proof of the convergence of the solution u^ε as $\varepsilon \to 0$ in the most critical case where $r_0 > 0$ and $\beta^0 > 0$, see Remark 2 for other cases. We also address, cf. Section 6, the asymptotic behavior or the eigenelements of the associated spectral problem

$$\int_\Omega \nabla u^\varepsilon . \nabla v \, dx + \beta(\varepsilon) \int_{\bigcup T^\varepsilon} \alpha \, u^\varepsilon v \, d\hat{x} = \lambda^\varepsilon \int_\Omega u^\varepsilon v \, dx, \ \forall v \in \mathbf{V}, \tag{10}$$

λ^ε being the spectral parameter and u^ε the corresponding eigenfunction.

For each fixed $\varepsilon > 0$, problem (10) has the discrete spectrum:

$$0 < \lambda_1^\varepsilon \leq \lambda_2^\varepsilon \leq \cdots \leq \lambda_n^\varepsilon \leq \cdots \xrightarrow{n \to \infty} +\infty, \tag{11}$$

where we have adopted the convention of repeated eigenvalues according to their multiplicities. We can choose the associated eigenfunctions $\{u_k^\varepsilon\}_{k=1}^\infty$ forming an orthonormal basis of $L^2(\Omega)$.

Using the Poincaré inequality and the minimax principle, for each fixed $n \in \mathbb{N}$, we show the following bounds for the eigenvalues:

$$0 < C \leq \lambda_n^\varepsilon \leq C_n \quad \forall \varepsilon > 0, \tag{12}$$

where C and C_n are constants independent of ε.

3 The homogenized problems

We use the technique of matched asymptotic expansions, following [11, 12], and [14] with the suitable modifications. For the sake of completeness, we outline here this technique. The general idea is to consider local and outer expansions along with matching principles as follows.

Assume an outer expansion for the solution of (6):

$$u^\varepsilon(x) = u^0(x) + \cdots, \quad \text{in } \Omega \cap \{x_3 > d\}, \quad \forall d > 0, \tag{13}$$

which is supposed to be valid out of certain "large" neighborhoods of the reaction regions $T_{\widetilde{x}_k}^\varepsilon$; by dots we indicate regular terms in the asymptotic series containing lower order functions of ε that we are not using in our analysis. Also, assume a local expansion in a neighborhood of each region $T_{\widetilde{x}_k}^\varepsilon$:

$$u^\varepsilon(x) = V^0(y) + \cdots \quad \text{for } y \in B(0, R) \cap \overline{\mathbb{R}^{3+}}, \quad \forall R > R_0, \tag{14}$$

R_0 such that $\overline{T} \subset B(0, R_0)$. Above, we denote by

$$y = \frac{x - \widetilde{x}_k}{r_\varepsilon} \tag{15}$$

the *local variable* in a neighborhood of each $\widetilde{x}_k \in \Sigma, \mathrm{k} \in \mathcal{J}^\varepsilon$.

By matching the local and outer expansions, at the first order, we can write

$$\lim_{|y| \to \infty} V^0(y) = \lim_{x \to \widetilde{x}_k} u^0(x). \tag{16}$$

By replacing (13) in (6), we obtain the following equations for $u^0(x)$:

$$-\Delta u^0 = f \text{ in } \Omega, \quad u^0 = 0 \text{ on } \Gamma_\Omega, \tag{17}$$

plus other boundary condition on Σ to be determined.

Similarly, taking derivatives with respect to y in (6), and using (14) and (16), we obtain the following equations for $V^0(y)$:

$$\begin{cases} -\Delta_y V^0 = 0 & \text{in } \mathbb{R}^{3+}, \\ \dfrac{\partial V^0}{\partial n_y} = 0 & \text{on } \{y_3 = 0\} \setminus T, \\ \dfrac{1}{r_\varepsilon} \dfrac{\partial V^0}{\partial n_y} + \beta(\varepsilon)\alpha(\widetilde{x}_k) V^0(y) = 0 & \text{on } T, \\ V^0(y) \to u^0(\widetilde{x}_k) & \text{as } |y| \to \infty, \quad y_3 > 0. \end{cases}$$

Thus, introducing a new function $W(y)$, we set

$$V^0(y) = u^0(\widetilde{x}_k)(1 - W(y)), \tag{18}$$

and depending on the value of β^0 in (3), we are led to different *local problems* satisfied by W (cf. (21) and (24)) and to $W \equiv 0$ when $\beta^0 = 0$ in (3).

Finally, to complete (17), depending on the different relations between ε, r_ε, and $\beta(\varepsilon)$ (cf. (1)–(3)), we use (13)–(18) and the well-known technique from the mechanics of continuous media of control volume in thin *coin-like domains* (see [11] for further details and references) to determine the boundary condition for u^0 on Σ. That is, we obtain the *homogenized problems*.

Local and homogenized problems are listed below:

- In the most critical situation when $\beta^0 > 0$ and $r_0 > 0$, the homogenized problem reads

$$\begin{cases} -\Delta_x u^0 = f & \text{in } \Omega, \\ u^0 = 0 & \text{on } \Gamma_\Omega, \\ \dfrac{\partial u^0}{\partial n_x} + r_0 C^e(x) u^0 = 0 \text{ on } \Sigma, \end{cases} \tag{19}$$

where C^e is the function defined as

$$C^e(x) = \int_T \frac{\partial W^{\alpha(x)}}{\partial n_y} \, d\hat{y}, \tag{20}$$

$W^{\alpha(x)}$ being the solution of the $\alpha(x)$-*dependent local problem* (cf. (18))

$$\begin{cases} -\Delta_y W^{\alpha(x)} = 0 & \text{in } \mathbb{R}^{3+}, \\ \dfrac{\partial W^{\alpha(x)}}{\partial n_y} = 0 & \text{on } \{y_3 = 0\} \setminus T, \\ \dfrac{\partial W^{\alpha(x)}}{\partial n_y} + \beta^0 \alpha(x)(W^{\alpha(x)} - 1) = 0 & \text{on } T, \\ W^{\alpha(x)}(y) \to 0 & \text{as } |y| \to \infty, \ y_3 > 0. \end{cases} \tag{21}$$

Above, and in what follows, the variable y denotes an auxiliary variable in $\mathbb{R}^3$, cf. (15), and the lower index x or y indicates the variable for derivatives, while the upper index $\alpha(x)$ refers to the parameter arising in the equation on T, which deals with the macroscopic variable x.

- For the critical size $r_0 > 0$, when $\beta^0 = +\infty$, the homogenized problem reads

$$\begin{cases} -\Delta_x u^0 = f & \text{in } \Omega, \\ u^0 = 0 & \text{on } \Gamma_\Omega, \\ \dfrac{\partial u^0}{\partial n_x} + r_0 C u^0 = 0 \text{ on } \Sigma, \end{cases} \tag{22}$$

where C is now a constant defined as

$$C = \left\langle \frac{\partial W}{\partial n_y}, 1 \right\rangle_{H^{-1/2}(T) \times H^{1/2}(T)}, \tag{23}$$

W being the solution of the *local problem* (cf. (18))

$$\begin{cases} -\Delta_y W = 0 & \text{in } \mathbb{R}^{3+}, \\ \dfrac{\partial W}{\partial n_y} = 0 & \text{on } \{y_3 = 0\} \setminus T, \\ W = 1 & \text{on } T, \\ W(y) \to 0 & \text{as } |y| \to \infty, \ y_3 > 0. \end{cases} \tag{24}$$

- For $\beta^* > 0$ and large sizes $r_0 = +\infty$, the homogenized problem reads

$$\begin{cases} -\Delta_x u^0 = f & \text{in } \Omega, \\ u^0 = 0 & \text{on } \Gamma_\Omega, \\ \dfrac{\partial u^0}{\partial n_x} + \beta^* |T| \alpha u^0 = 0 & \text{on } \Sigma. \end{cases} \tag{25}$$

- For the extreme cases where $\beta^* = 0$ or $r_0 = 0$, the homogenized problem is

$$\begin{cases} -\Delta_x u^0 = f & \text{in } \Omega, \\ u^0 = 0 & \text{on } \Gamma_\Omega, \\ \dfrac{\partial u^0}{\partial n_x} = 0 & \text{on } \Sigma. \end{cases} \tag{26}$$

- For the extreme cases where $\beta^0 > 0$ or $\beta^0 = +\infty$ and $r_0 = +\infty$, the homogenized problem is the Dirichlet problem

$$\begin{cases} -\Delta_x u^0 = f & \text{in } \Omega, \\ u^0 = 0 & \text{on } \partial\Omega. \end{cases} \tag{27}$$

The variational formulations of (27) in $H_0^1(\Omega)$ and of (26) and (25) in $\mathbf{V}$ are classical in the literature. The existence of a unique solution of the Robin problems (22) and (19) in $\mathbf{V}$ follows from the positivity of the capacities C and C^e, respectively, and the continuity of C^e, $C^e \in C(\overline{\Sigma})$ (cf. Proposition 1). Indeed, for problem (22), the positivity of C is well known in the literature (see, e.g., [19, 21], and [26]). Regarding problem (19), we show the positivity of the function C^e in (20) from the properties of the solution of the local problem (21) (see (33)):

$$C^e(x) = \int_{\mathbb{R}^{3+}} |\nabla_y W^{\alpha(x)}|^2 dy + \beta^0 \alpha(x) \int_T (W^{\alpha(x)}(y) - 1)^2 d\hat{y} > 0.$$

In fact, excepting problem (27), the rest of homogenized problems have the weak formulation: Find $u^0 \in \mathbf{V}$ satisfying

$$\int_\Omega \nabla_x u^0 . \nabla_x v \, dx + \int_\Sigma M(\hat{x}) u^0 v \, d\hat{x} = \int_\Omega f v \, dx, \quad \forall v \in \mathbf{V}, \tag{28}$$

where $M(\hat{x})$ is defined as the function $r_0 C^e(\hat{x})$ or $\beta^0 |T| \alpha(\hat{x})$ when dealing with problem (19) or (25), respectively, and it is defined as the constant $r_0 C$ or 0 when dealing with problem (22) or (26).

Similarly, the corresponding homogenized spectral problems read: Find $\lambda^0, u^0 \in \mathbf{V}, u^0 \neq 0$, satisfying

$$\int_\Omega \nabla_x u^0 . \nabla_x v \, dx + \int_\Sigma M(\hat{x}) u^0 v \, d\hat{x} = \lambda^0 \int_\Omega u^0 v \, dx, \quad \forall v \in \mathbf{V}. \tag{29}$$

They have the discrete spectrum

$$0 < \lambda_1^0 \leq \lambda_2^0 \leq \cdots \leq \lambda_n^0 \leq \cdots \xrightarrow{n \to \infty} +\infty, \tag{30}$$

where, also, we have adopted the convention of repeated eigenvalues according to their multiplicities.

In Sections 5 and 6, we show the convergence of solutions and eigenelements when $M(\hat{x}) = r_0 C^e(\hat{x})$.

4 The solutions of the local problems

The setting of the local problem (24) in the suitable Sobolev space and the behavior at infinity of the solution has been considered in [26] (cf. also [19, 21] and Remark 1). Here, we deal with the abstract framework of the parameter family of local problems (21) and the properties of their solutions, the parameter being $\alpha(x)$ for $x \in \overline{\Omega}$.

Consider the space $\mathcal{V}$ completion of $\mathcal{D}(\overline{\mathbb{R}^{3+}})$ with respect to the norm

$$\|U\|_{\mathcal{V}} = \left(\|\nabla_y U\|_{L^2(\mathbb{R}^{3+})}^2 + \|U\|_{L^2(T)}^2 \right)^{1/2}.$$

For each $x \in \overline{\Omega}$, let us define the bilinear, symmetric, continuous, and coercive form on $\mathcal{V}$:

$$a_x(U, V) = \int_{\mathbb{R}^{3+}} \nabla_y U . \nabla_y V \, dy + \beta^0 \alpha(x) \int_T UV \, d\hat{y}, \quad \forall U, V \in \mathcal{V},$$

which defines a norm in $\mathcal{V}$ equivalent to $\| \cdot \|_{\mathcal{V}}$. Also, we consider the continuous functional on $\mathcal{V}$

$$F_x(U) = \beta^0 \alpha(x) \int_T U \, d\hat{y}, \quad \forall U \in \mathcal{V}.$$

Then, the Riesz representation theorem ensures that there is a unique function $W^{\alpha(x)} \in \mathcal{V}$ satisfying

$$a_x(W^{\alpha(x)}, V) = F_x(V), \quad \forall V \in \mathcal{V}, \tag{31}$$

which is the weak formulation of (21). In order to be more precise with the behavior at infinity of this solution, we use, for instance, the technique in [11] (cf. also [19] and Appendix in [26]).

Theorem 1 *For fixed* $x \in \overline{\Omega}$, *the unique solution* $W^{\alpha(x)}$ *of* (31) *admits the representation*

$$W^{\alpha(x)}(y) = \frac{\mathcal{K}(x)}{|y|} + O\Big(\frac{1}{|y|^2}\Big), \quad as \; |y| \to \infty, \tag{32}$$

where $\mathcal{K}(x)$ *is defined by the following chain of equalities:*

$$2\pi \mathcal{K}(x) = \int_T \frac{\partial W^{\alpha(x)}}{\partial n_y} d\hat{y} = \beta^0 \alpha(x) \int_T (1 - W^{\alpha(x)}(y)) d\hat{y}$$
$$= \int_{\mathbb{R}^{3+}} |\nabla_y W^{\alpha(x)}|^2 dy + \beta^0 \alpha(x) \int_T (W^{\alpha(x)}(y) - 1)^2 d\hat{y}. \tag{33}$$

Proof The behavior at infinity of the solution of (31) is obtained, for instance, as a consequence of the fact that $W^{\alpha(x)}$ can be extended to a harmonic function W^x outside a ball $B(0, R_0)$ such that $\overline{T} \subset B(0, R_0)$, with $\nabla_y W^x \in L^2(\mathbb{R}^3 \setminus B(0, R_0))$. The representation of W^x in $L^2(\partial B(0, R_0))$ in terms of the spherical harmonics leads to (32): see Section II in [17] and Section IV.8 in [28].

Let us show the chain of equalities (33). Considering the first equation in (21), the first equality in (33) is obtained by applying the Green formula in $B(0, R) \cap \mathbb{R}^{3+}$, with $B(0, R)$ any ball of radius $R > R_0$. We have

$$\int_T \frac{\partial W^{\alpha(x)}}{\partial n_y} d\hat{y} = - \int_{\partial B(0,R) \cap \mathbb{R}^{3+}} \frac{\partial W^{\alpha(x)}}{\partial \nu_y} ds_y,$$

ν_y being the unit outer normal to $\partial B(0, R)$. Taking limits when $R \to \infty$ and (32) give

$$\int_T \frac{\partial W^{\alpha(x)}}{\partial n_y} d\hat{y} = \mathcal{K}(x) 2\pi.$$

The boundary condition on T in (21) gives the second equality in (33), while taking $V = W^{\alpha(x)}$ in (31) provides the last equality in (33). Hence, the theorem holds. □

As a consequence of Theorem 1, in the following proposition, we show that the solution $W^{\alpha(x)}$ of (21) and certain other related functions are continuous functions of the parameter $x \in \overline{\Omega}$.

Proposition 1 *The solution $W^{\alpha(x)}$ of (21) depends continuously on $x \in \overline{\Omega}$ in the topology of $\mathcal{V}$. In addition, the functions*

$$\int_{\mathbb{R}^{3+}} |\nabla_y W^{\alpha(x)}|^2 dy, \quad \int_T (W^{\alpha(x)})^2 d\hat{y}, \quad \int_T W^{\alpha(x)} d\hat{y}, \quad \text{and} \quad \mathcal{K}(x)$$

depend continuously on $x \in \overline{\Omega}$. Also, the distributional partial derivatives of $\mathcal{K}$ satisfy

$$\frac{\partial \mathcal{K}}{\partial x_i} \in L^{\infty}(\Omega), \quad i = 1, 2, 3. \tag{34}$$

Proof First, let us note that writing $V = W^{\alpha(x)}$ in (31), we get

$$\int_{\mathbb{R}^{3+}} |\nabla_y W^{\alpha(x)}|^2 \, dy + \beta^0 \alpha(x) \int_T (W^{\alpha(x)})^2 d\hat{y} = \beta^0 \alpha(x) \int_T W^{\alpha(x)} \, d\hat{y},$$

and applying the Cauchy–Schwarz inequality, we have the bounds

$$\|W^{\alpha(x)}\|_{L^2(T)} \leq C \quad \text{and} \quad \|\nabla_y W^{\alpha(x)}\|_{L^2(\mathbb{R}^{3+})} \leq C, \tag{35}$$

where, here, and throughout the proof, C denotes a constant independent of x.

Let us show the continuity in the statement of the proposition. In order to do this, for $x, x' \in \overline{\Omega}$, let us consider (31) for $V = W^{\alpha(x)} - W^{\alpha(x')}$ and the weak formulation for $W^{\alpha(x')}$ and $V = W^{\alpha(x)} - W^{\alpha(x')}$ and subtract both.
We obtain

$$\int_{\mathbb{R}^{3+}} |\nabla_y (W^{\alpha(x)} - W^{\alpha(x')})|^2 \, dy + \beta^0 \alpha(x) \int_T (W^{\alpha(x)} - W^{\alpha(x')})^2 d\hat{y}$$

$$= \beta^0 |\alpha(x) - \alpha(x')| \left| \int_T W^{\alpha(x')} (W^{\alpha(x)} - W^{\alpha(x')}) d\hat{y} + \int_T (W^{\alpha(x)} - W^{\alpha(x')}) \, d\hat{y} \right|$$

$$\leq |\alpha(x) - \alpha(x')| \, \beta^0 \left(\|W^{\alpha(x')}\|_{L^2(T)} + |T|^{1/2} \right) \|W^{\alpha(x)} - W^{\alpha(x')}\|_{L^2(T)}.$$

From (35), we can write

$$\|W^{\alpha(x)} - W^{\alpha(x')}\|_{\mathcal{V}} \leq C \, |\alpha(x) - \alpha(x')|$$

and

$$\|W^{\alpha(x)} - W^{\alpha(x')}\|_{L^2(T)} \le C \left|\alpha(x) - \alpha(x')\right|,$$

which imply the continuity of the two first integrals in the statement of the proposition. The continuity of the third integral is obtained on account of the last inequality above and the Cauchy–Schwarz inequality. That is,

$$\left| \int_T (W^{\alpha(x)} - W^{\alpha(x')}) d\hat{y} \right| \le C \left|\alpha(x) - \alpha(x')\right|.$$

Finally, we show the continuity for $\mathcal{K}$ taking into account its definitions, (33), (35), and the above inequality. Indeed, we have

$$
|\mathcal{K}(x) - \mathcal{K}(x')| = \frac{\beta^0}{2\pi} \left|\alpha(x) \int_T (1 - W^{\alpha(x)}(y)) d\hat{y} - \alpha(x') \int_T (1 - W^{\alpha(x')}(y)) d\hat{y}\right|
$$
$$
\le C|\alpha(x) - \alpha(x')|.
$$

Hence, the continuity of the function α gives the continuity of $\mathcal{K}$. As regards the boundedness for the distributional partial derivatives of $\mathcal{K}$, it is consequence of the last inequality and property (5) for $\alpha(x)$, which ensures that $\mathcal{K}$ is a uniformly Lipschitz continuous function; namely,

$$|\mathcal{K}(x) - \mathcal{K}(x')| \le K|x - x'|, \quad \forall x, x' \in \overline{\Omega},$$

with K a constant independent of x and x', and (34) also holds, see Sections III.24 and III.28 in [30]. Thus, the proposition is proved.   □

Proposition 2 *For $x \in \overline{\Omega}$, the solution $W^{\alpha(x)}$ of (31) satisfies*

$$\left|W^{\alpha(x)}(y)\right| \le \frac{C}{d(y, \overline{T})} \quad and \quad \left|\frac{\partial W^{\alpha(x)}}{\partial y_j}(y)\right| \le \frac{C}{d(y, \overline{T})^2}, \quad \forall y \in \mathbb{R}^{3+}, \quad (36)$$

for $j = 1, 2, 3$, where C is a constant independent of x.

Proof Denoting by

$$q^x := \frac{\partial W^{\alpha(x)}}{\partial n_y}\bigg|_T,$$

a solution of (21) reads

$$W^{\alpha(x)}(y) = -\frac{1}{2\pi} \int_T \frac{1}{\sqrt{(y_1 - \xi_1)^2 + (y_2 - \xi_2)^2 + y_3^2}} q^x(\xi_1, \xi_2) d\hat{\xi}, \quad \forall y \in \mathbb{R}^{3+}$$

(cf., e.g., [16, 26] and [29]). To show that this function belongs to $\mathcal{V}$, we follow the technique in Theorem 4.1 in [19] with minor modifications.

As a consequence, we have (36) for $C \equiv C_x$ a certain constant, which in principle can depend on x; namely,

$$C_x = \frac{1}{2\pi} \int_T |q^x(\hat{\xi})| \, d\hat{\xi}.$$

Taking into account the equation on T in (21), the continuity of $\alpha(x)$ and $\|W^{\alpha(x)}\|_{L^2(T)}$, cf. Proposition 1, the Cauchy–Schwarz inequality allows us to derive that C_x is bounded independently of x, and the result of the proposition holds. □

Remark 1 In connection with the properties of the solution of (24), they follow the scheme above for the solution of problem (21) with the suitable modifications (see the Appendix in [26] for details). Let us mention that the space of the setting of the problem is the space completion of $\mathcal{D}(\overline{\mathbb{R}^{3+}})$ with respect to the Dirichlet norm, and that the constant arising in (23) is also given by

$$C = \int_{\mathbb{R}^{3+}} |\nabla_y W|^2 \, dy.$$

4.1 The test functions in the most critical case

In this section, we introduce some auxiliary functions that allow us to construct the test functions (cf. (44)) to prove the convergence when $r_0 > 0$ and $\beta^0 > 0$.

Let us consider $\varphi \in C^\infty[0, 1]$, $0 \le \varphi \le 1$, $\varphi = 1$ in $[0, 1/8]$ and $\mathrm{Supp}(\varphi) \subset [0, 1/4]$, we construct the function

$$\varphi^\varepsilon(x) = \begin{cases} 1 & \text{for } x \in \bigcup_{k \in \mathcal{J}^\varepsilon} \overline{B^+\left(\widetilde{x}_k, r_\varepsilon + \frac{\varepsilon}{8}\right)}, \\ \varphi\left(\dfrac{|x - \widetilde{x}_k| - r_\varepsilon}{\varepsilon}\right) & \text{for } x \in \mathcal{C}_{\widetilde{x}_k}^{\varepsilon,+}, k \in \mathcal{J}^\varepsilon, \\ 0 & \text{for } x \in \Omega \setminus \bigcup_{k \in \mathcal{J}^\varepsilon} B^+\left(\widetilde{x}_k, r_\varepsilon + \frac{\varepsilon}{4}\right), \end{cases} \tag{37}$$

where $\mathcal{J}^\varepsilon = \{k \in \mathbb{Z}^2 : T_{\widetilde{x}_k}^\varepsilon \subset \Sigma\}$, $B^+(\widetilde{x}_k, r)$ denotes the half-ball of radius r centered at the point $\widetilde{x}_k$, namely, $B(\widetilde{x}_k, r) \cap \{x_3 > 0\}$, and $\mathcal{C}_{\widetilde{x}_k}^{\varepsilon,+}$ stands for the half-annulus

$$\mathcal{C}_{\widetilde{x}_k}^{\varepsilon,+} = B^+\left(\widetilde{x}_k, r_\varepsilon + \frac{\varepsilon}{4}\right) \setminus B^+\left(\widetilde{x}_k, r_\varepsilon + \frac{\varepsilon}{8}\right).$$

Let us define the functions $\widetilde{W}^\varepsilon(x)$, which we construct from the solutions of the local problems (21), as follows: we set

$$W^{k,\varepsilon}(x) = W^{\alpha(\widetilde{x}_k)}\left(\frac{x - \widetilde{x}_k}{r_\varepsilon}\right)\varphi^\varepsilon(x) \quad \text{for} \quad x \in B^+\left(\widetilde{x}_k, r_\varepsilon + \frac{\varepsilon}{4}\right),$$

and

$$\widetilde{W}^{k,\varepsilon}(x) = 1 - W^{k,\varepsilon}(x) \quad \text{for} \quad x \in B^+\left(\widetilde{x}_k, r_\varepsilon + \frac{\varepsilon}{4}\right),$$

which we extend by 1 in $\Omega \setminus \bigcup B^+\left(\widetilde{x}_k, r_\varepsilon + \frac{\varepsilon}{4}\right)$. Then,

$$\widetilde{W}^\varepsilon(x) = \begin{cases} \widetilde{W}^{k,\varepsilon}(x) & \text{for } x \in B^+\left(\widetilde{x}_k, r_\varepsilon + \frac{\varepsilon}{4}\right), \quad k \in \mathcal{J}^\varepsilon, \\ 1 & \text{for } x \in \Omega \setminus \bigcup B^+\left(\widetilde{x}_k, r_\varepsilon + \frac{\varepsilon}{4}\right). \end{cases} \tag{38}$$

Proposition 3 *There is a constant C independent of ε such that $\forall x \in \mathcal{C}_{\widetilde{x}_k}^{\varepsilon,+}$, and we have*

$$\left|\frac{\partial \varphi^\varepsilon}{\partial x_j}(x)\right| \le C\frac{1}{\varepsilon}, \quad j = 1, 2, 3, \tag{39}$$

$$\left|W^{\alpha(\widetilde{x}_k)}\left(\frac{x - \widetilde{x}_k}{r_\varepsilon}\right)\right| \le C\frac{r_\varepsilon}{\varepsilon}, \quad \left|\frac{\partial W^{\alpha(\widetilde{x}_k)}}{\partial x_j}\left(\frac{x - \widetilde{x}_k}{r_\varepsilon}\right)\right| \le C\frac{r_\varepsilon}{\varepsilon^2}, \quad j = 1, 2, 3, \tag{40}$$

and

$$|W^{k,\varepsilon}(x)| \le C\frac{r_\varepsilon}{\varepsilon}, \quad \left|\frac{\partial W^{k,\varepsilon}}{\partial x_j}(x)\right| \le C\frac{r_\varepsilon}{\varepsilon^2}, \quad j = 1, 2, 3. \tag{41}$$

In addition,

$$\|\widetilde{W}^\varepsilon\|_{H^1(\Omega)} \le C \quad \text{and} \quad \widetilde{W}^\varepsilon \xrightarrow{\varepsilon \to 0} 1 \quad \text{in } H^1(\Omega)\text{-weak.} \tag{42}$$

Proof Bound (39) is a consequence of the definition (37), while bounds (40) are a consequence of (36). Estimates (39) and (40) give (41). Let us show (42).

First, using (41), (15), (35), (4), and $r_0 > 0$ in (2), we have

$$\|\nabla \widetilde{W}^\varepsilon\|_{L^2(\Omega)}^2 = \sum_{\widetilde{x}_k} \left\|\nabla_x W^{k,\varepsilon}\right\|_{L^2(\mathcal{C}_{\widetilde{x}_k}^{\varepsilon,+})}^2 + \sum_{\widetilde{x}_k} \left\|\nabla_x W^{k,\varepsilon}\right\|_{L^2(B^+(\widetilde{x}_k, r_\varepsilon + \varepsilon/8))}^2$$

$$\le C\frac{r_\varepsilon^2}{\varepsilon^4} \sum_{\widetilde{x}_k} \int_{\mathcal{C}_{\widetilde{x}_k}^{\varepsilon,+}} dx + r_\varepsilon \sum_{\widetilde{x}_k} \left\|\nabla_y W^{\alpha(\widetilde{x}_k)}\right\|_{L^2(B^+(0, 1 + \varepsilon/(r_\varepsilon 8)))}^2 \le C. \tag{43}$$

Then, we obtain the following estimate:

$$\left\| \widetilde{W}^\varepsilon - 1 \right\|_{L^2(\Omega)}^2 = \sum_{\widetilde{x}_k} \left\| W^{k,\varepsilon} \right\|_{L^2(B^+(\widetilde{x}_k, r_\varepsilon + \varepsilon/4))}^2$$

$$\leq C\varepsilon^2 \sum_{\widetilde{x}_k} \left\| \nabla_x W^{k,\varepsilon} \right\|_{L^2(B^+(\widetilde{x}_k, r_\varepsilon + \varepsilon/4))}^2$$

$$\leq \varepsilon^2 C \left\| \nabla_x \widetilde{W}^\varepsilon \right\|_{L^2(\Omega)}^2 \leq C\varepsilon^2,$$

where we have used the definition (37), the Poincaré inequality on each half-ball, and (43). Thus, the convergence of $(\widetilde{W}^\varepsilon - 1)$ toward zero in $(L^2(\Omega))^3$ holds, as $\varepsilon \to 0$, and so does the bound in (42). This concludes the proof of the proposition.

$\square$

5 The convergence

In this section, considering $r_0 > 0$ and $\beta^0 > 0$, we show that the limit of u^ε in $H^1(\Omega)$-weak is the solution of the homogenized problem (19), cf. Theorem 2. For convenience, first, we introduce a convergence result of measures (see [18] for the proof).

Lemma 1 *Let us consider $a_\varepsilon < \varepsilon$ such that $a_\varepsilon \varepsilon^{-1} \to a_0$, and let $B(\widetilde{x}_k, a_\varepsilon)$ denote the ball of radius a_ε centered at $\widetilde{x}_k$. Then, $\forall w \in H_0^1(\Omega)$,*

$$\left| \sum_{\widetilde{x}_k} \int_{\partial B(\widetilde{x}_k, a_\varepsilon)} w \, ds_x - 4\pi a_0^2 \int_\Sigma w \, d\hat{x} \right| \leq C(\varepsilon^{1/2} + |a_\varepsilon \varepsilon^{-1} - a_0|) \|w\|_{H^1(\Omega)}.$$

Theorem 2 *Let us consider the case where $r_0 > 0$ and $\beta^0 > 0$ in (2) and (3), respectively. Then, as $\varepsilon \to 0$, the solution of (7) converges in $H^1(\Omega)$-weak toward the solution of (28) for $M(\hat{x}) = r_0 C^e(\hat{x})$.*

Proof Let us consider $\phi \in \{v \in C^1(\overline{\Omega}) : v = 0 \text{ on } \Gamma_\Omega\}$. From the definition of the spaces $\mathcal{V}$ and $\mathbf{V}$, $W^{\alpha(\hat{x})}$, (38), and (42), we have that $\phi \widetilde{W}^\varepsilon \in \mathbf{V}$ and $\phi \widetilde{W}^\varepsilon \to \phi$ in $H^1(\Omega)$-weak as $\varepsilon \to 0$. Then, we write (7) for the test function $v = \phi \widetilde{W}^\varepsilon$

$$\int_\Omega \nabla u^\varepsilon . \nabla(\phi \widetilde{W}^\varepsilon) \, dx + \beta(\varepsilon) \int_{\bigcup T^\varepsilon} \alpha \, u^\varepsilon \phi \widetilde{W}^\varepsilon \, d\hat{x} = \int_\Omega f \phi \widetilde{W}^\varepsilon \, dx. \qquad (44)$$

This amounts to

$$\int_\Omega \nabla u^\varepsilon . \nabla \phi \widetilde{W}^\varepsilon \, dx + \int_\Omega \nabla(u^\varepsilon \phi) . \nabla \widetilde{W}^\varepsilon \, dx - \int_\Omega u^\varepsilon \nabla \phi . \nabla \widetilde{W}^\varepsilon \, dx$$

$$+ \beta(\varepsilon) \int_{\bigcup T^\varepsilon} \alpha \, u^\varepsilon \phi \widetilde{W}^\varepsilon \, d\hat{x} = \int_\Omega f \phi \widetilde{W}^\varepsilon \, dx.$$

Taking limits as $\varepsilon \to 0$, we get

$$
\begin{aligned}
&\int_{\Omega} \nabla u^0 . \nabla \phi \, dx - \int_{\Omega} f \phi \, dx \\
&= - \lim_{\varepsilon \to 0} \left(\int_{\Omega} \nabla (u^{\varepsilon} \phi) . \nabla \widetilde{W}^{\varepsilon} \, dx + \beta(\varepsilon) \int_{\bigcup T^{\varepsilon}} \alpha \, u^{\varepsilon} \phi \widetilde{W}^{\varepsilon} \, d\hat{x} \right).
\end{aligned}
\tag{45}
$$

Let us compute the limit on the right-hand side of (45).

For brevity, in connection with the integrals on $B^{+}(\widetilde{x}_k, r_{\varepsilon} + \frac{\varepsilon}{8})$, we introduce the following notations:

$$
W^{k,r_{\varepsilon}}(x) \equiv W^{\alpha(\widetilde{x}_k)} \left(\frac{x - \widetilde{x}_k}{r_{\varepsilon}} \right) \quad \text{and} \quad \Gamma_{\widetilde{x}_k, r_{\varepsilon} + \frac{\varepsilon}{8}} = \partial B(\widetilde{x}_k, r_{\varepsilon} + \frac{\varepsilon}{8}) \cap \mathbb{R}^{3+}.
$$

Thus, for the first integral inside the limit, we have

$$
\begin{aligned}
\int_{\Omega} \nabla (u^{\varepsilon} \phi) . \nabla \widetilde{W}^{\varepsilon} \, dx &= - \sum_{\widetilde{x}_k} \int_{B^{+}(\widetilde{x}_k, r_{\varepsilon} + \frac{\varepsilon}{8})} \nabla_x (u^{\varepsilon} \phi) . \nabla_x W^{k,r_{\varepsilon}} \, dx \\
&\quad - \sum_{\widetilde{x}_k} \int_{\mathscr{C}^{\varepsilon,+}_{\widetilde{x}_k}} \nabla_x (u^{\varepsilon} \phi) . \nabla_x W^{k,r_{\varepsilon}} \varphi^{\varepsilon} \, dx - \sum_{\widetilde{x}_k} \int_{\mathscr{C}^{\varepsilon,+}_{\widetilde{x}_k}} \nabla_x (u^{\varepsilon} \phi) . \nabla_x \varphi^{\varepsilon} \, W^{k,r_{\varepsilon}} \, dx.
\end{aligned}
$$

Taking into account estimates in Proposition 3, (8), (4), (2), and the measure of each $\mathscr{C}^{\varepsilon,+}_{\widetilde{x}_k}$, and applying the Cauchy–Schwarz inequality, we show that the two last sums of integrals above are bounded by $C\sqrt{\varepsilon}$. Consequently, using the Green formula in $B^{+}(\widetilde{x}_k, r_{\varepsilon} + \frac{\varepsilon}{8})$, the limit on the right-hand side of (45) is given by

$$
\begin{aligned}
\mathbf{L} &= - \lim_{\varepsilon \to 0} \sum_{\widetilde{x}_k} \int_{B^{+}(\widetilde{x}_k, r_{\varepsilon} + \frac{\varepsilon}{8})} \nabla_x (u^{\varepsilon} \phi) . \nabla_x W^{k,r_{\varepsilon}} \, dx \\
&\quad + \lim_{\varepsilon \to 0} \beta(\varepsilon) \sum_{\widetilde{x}_k} \int_{T^{\varepsilon}_{\widetilde{x}_k}} \alpha \, u^{\varepsilon} \phi (1 - W^{k,r_{\varepsilon}}) \, d\hat{x} \\
&= - \lim_{\varepsilon \to 0} \sum_{\widetilde{x}_k} \int_{\Gamma_{\widetilde{x}_k, r_{\varepsilon} + \frac{\varepsilon}{8}}} u^{\varepsilon} \phi \frac{\partial W^{k,r_{\varepsilon}}}{\partial \nu_x} \, ds_x - \lim_{\varepsilon \to 0} \sum_{\widetilde{x}_k} \int_{T^{\varepsilon}_{\widetilde{x}_k}} u^{\varepsilon} \phi \frac{\partial W^{k,r_{\varepsilon}}}{\partial n_x} \, d\hat{x} \\
&\quad + \lim_{\varepsilon \to 0} \sum_{\widetilde{x}_k} \beta(\varepsilon) \int_{T^{\varepsilon}_{\widetilde{x}_k}} \alpha \, u^{\varepsilon} \phi (1 - W^{k,r_{\varepsilon}}) \, d\hat{x}.
\end{aligned}
$$

Writing the boundary condition for $W^{\alpha(\hat{x})}$ on T in the macroscopic variable, cf. (21) and (15), we show that

$$\mathbf{L} = -\lim_{\varepsilon \to 0} \sum_{\widetilde{x}_k} \int_{\Gamma_{\widetilde{x}_k, r_\varepsilon + \frac{\varepsilon}{8}}} u^\varepsilon \phi \frac{\partial W^{k, r_\varepsilon}}{\partial \nu_x} ds_x.$$

Indeed, on account of (3), (2), (4), (5), (8), and Proposition 1, we deduce

$$-\sum_{\widetilde{x}_k} \int_{T^\varepsilon_{\widetilde{x}_k}} u^\varepsilon \phi \frac{\partial W^{k, r_\varepsilon}}{\partial n_x} d\hat{x} + \sum_{\widetilde{x}_k} \beta(\varepsilon) \int_{T^\varepsilon_{\widetilde{x}_k}} \alpha \, u^\varepsilon \phi (1 - W^{k, r_\varepsilon}) \, d\hat{x}$$

$$= -\sum_{\widetilde{x}_k} \frac{1}{r_\varepsilon} \beta^0 \alpha(\widetilde{x}_k) \int_{T^\varepsilon_{\widetilde{x}_k}} u^\varepsilon \phi (1 - W^{k, r_\varepsilon}) \, d\hat{x}$$

$$+ \sum_{\widetilde{x}_k} \beta(\varepsilon) \int_{T^\varepsilon_{\widetilde{x}_k}} \alpha \, u^\varepsilon \phi (1 - W^{k, r_\varepsilon}) \, d\hat{x}$$

$$= -\sum_{\widetilde{x}_k} \beta(\varepsilon) \alpha(\widetilde{x}_k) \int_{T^\varepsilon_{\widetilde{x}_k}} u^\varepsilon \phi (1 - W^{k, r_\varepsilon}) \, d\hat{x} + o_\varepsilon(1)$$

$$+ \sum_{\widetilde{x}_k} \beta(\varepsilon) \int_{T^\varepsilon_{\widetilde{x}_k}} \alpha \, u^\varepsilon \phi (1 - W^{k, r_\varepsilon}) \, d\hat{x}$$

$$= \sum_{\widetilde{x}_k} \beta(\varepsilon) \int_{T^\varepsilon_{\widetilde{x}_k}} (\alpha(x) - \alpha(\widetilde{x}_k)) \, u^\varepsilon \phi (1 - W^{k, r_\varepsilon}) \, d\hat{x} + o_\varepsilon(1) = o_\varepsilon(1),$$

where we have denoted by $o_\varepsilon(1)$ terms such that $o_\varepsilon(1) \to 0$, as $\varepsilon \to 0$.

In order to compute $\mathbf{L}$, we perform the change of variable from x to y, cf. (15), and use the representation (32). Then, we can write

$$\mathbf{L} = \lim_{\varepsilon \to 0} \frac{r_\varepsilon}{(r_\varepsilon + (\varepsilon/8))^2} \sum_{\widetilde{x}_k} \int_{\Gamma_{\widetilde{x}_k, r_\varepsilon + \frac{\varepsilon}{8}}} \mathcal{K}(\widetilde{x}_k) u^\varepsilon \phi \, ds_x$$

$$= \lim_{\varepsilon \to 0} \frac{r_\varepsilon}{(r_\varepsilon + (\varepsilon/8))^2} \sum_{\widetilde{x}_k} \int_{\Gamma_{\widetilde{x}_k, r_\varepsilon + \frac{\varepsilon}{8}}} \mathcal{K}(x) u^\varepsilon \phi \, ds_x.$$

Above, we have used (2), (5), Proposition 1, and the fact that

$$\sum_{\widetilde{x}_k} \int_{\Gamma_{\widetilde{x}_k, r_\varepsilon + \frac{\varepsilon}{8}}} |u^\varepsilon \phi| \, dx \leq C,$$

which holds because the sequence $|u^\varepsilon \phi|$ is bounded in $H^1(\Omega)$ (cf., e.g., III.28.I in [30]) and Lemma 1.

Now, we use the smoothness of $\mathcal{K}$ in Proposition 1, which guaranties that $\mathcal{K}u^\varepsilon\phi \in H^1(\Omega)$, and considering (8) gives

$$\|\mathcal{K}\phi u^\varepsilon\|_{H^1(\Omega)} \leq C.$$

Hence, extending by symmetry these functions to the lower half-space $\{x_3 < 0\}$, we get a sequence of functions $\widehat{\mathcal{K}\phi u^\varepsilon} \in H^1_0(\widehat{\Omega})$, satisfying

$$\widehat{\mathcal{K}\phi u^\varepsilon} \to \widehat{\mathcal{K}\phi u^0} \quad \text{in } H^1(\widehat{\Omega}) - weak, \quad \text{as } \varepsilon \to 0,$$

where by $\widehat{\Omega}$ we denote the extended domain of Ω by symmetry. Therefore, using this extension and Lemma 1, we have

$$\mathbf{L} = 8^2 r_0 \lim_{\varepsilon \to 0} \sum_{\tilde{x}_k} \int_{\Gamma_{\tilde{x}_k, r_\varepsilon + \frac{\varepsilon}{8}}} \mathcal{K}u^\varepsilon\phi \, ds_x$$

$$= 8^2 r_0 \frac{1}{2} \lim_{\varepsilon \to 0} \sum_{\tilde{x}_k} \int_{\partial B^+(\tilde{x}_k, r_\varepsilon + \frac{\varepsilon}{8})} \widehat{\mathcal{K}u^\varepsilon}\phi \, dx = r_0 2\pi \int_\Sigma \mathcal{K}u^0\phi \, d\hat{x}.$$

Finally, from (20) and formula (33), we obtain

$$\mathbf{L} = r_0 \int_\Sigma C^e u^0\phi d\hat{x}.$$

Consequently, we have computed the right-hand side of (45), namely $-\mathbf{L}$, and therefore u^0 satisfies

$$\int_\Omega \nabla u^0.\nabla\phi \, dx + r_0 \int_\Sigma C^e u^0\phi \, d\hat{x} = \int_\Omega f\phi \, dx, \quad \forall\phi \in C^1(\overline{\Omega}) : \phi = 0 \text{ on } \Gamma_\Omega,$$

and by density,

$$\int_\Omega \nabla u^0.\nabla v \, dx + r_0 \int_\Sigma C^e u^0 v \, d\hat{x} = \int_\Omega f v \, dx, \quad \forall v \in \mathbf{V},$$

which is the variational formulation of (19). Thus, the theorem is proved.   □

Remark 2 As regards the convergence of solutions in the rest of the cases stated in Section 3, it should be noted that, when $r_0 = 0$, the convergence (42) takes place in $H^1(\Omega)$ (cf. (43)), and the proof above simplifies providing that u^0 in (9) is the solution of (26).

In the case where $r_0 > 0$ and $\beta^0 = +\infty$ in (2) and (3), respectively, the test functions are constructed by replacing the solution of the local problem (21) by that of (24) (cf. Remark 1). Computing the limit in the right-hand side of (45) and obtaining that u^0 in (9) is the solution of (22) will likely follow the technique in [26] (cf. also [4] and [20]). Similarly, the idea applies when $r_0 = 0$ and $\beta^0 = +\infty$. Let us refer to the technique in [25] when $r_0 = +\infty$.

6 The spectral convergence

Using Theorem 2 and a result from the spectral perturbation theory, we show the convergence of the eigenvalues λ^ε of (10) toward those of (29) with conservation of multiplicity. For the sake of completeness, in Lemma 2, we introduce a simplified version of the abovementioned result, and we refer to Lemma 1.6 in Chapter III of [24] for the proof. Also, considering the bound (12), we mention Section III.9.1 in [2] for an alternative technique that can be applied. See [8, 9] and [10] for spectral problems for the Laplacian in perforated domains by balls with large parameters on the boundary.

Lemma 2 *Let $\mathbf{H}$ be a separable Hilbert space with the norm $\| \cdot \|$. Let $\mathcal{A}^\varepsilon$, $\mathcal{A}^0 \in \mathcal{L}(\mathbf{H})$ and $\mathcal{W}$ be a subspace of $\mathbf{H}$ such that $Im\, \mathcal{A}^0 = \{v \mid v = \mathcal{A}^0 u : u \in \mathbf{H}\} \subset \mathcal{W}$. We assume that the following properties are satisfied:*

i1).   *$\mathcal{A}^\varepsilon$ and $\mathcal{A}^0$ are positive, compact, and self-adjoint operators on $\mathbf{H}$ and $\|\mathcal{A}^\varepsilon\|_{\mathcal{L}(\mathbf{H})} \leq \mathbf{c}$, where $\mathbf{c}$ denotes a constant independent of ε.*

i2).   *For any $f \in \mathcal{W}$, $\|\mathcal{A}^\varepsilon f - \mathcal{A}^0 f\| \to 0$ as $\varepsilon \to 0$.*

i3).   *The family of operators $\mathcal{A}^\varepsilon$ is uniformly compact, which is to say, for any sequence $f^\varepsilon \in \mathbf{H}$ such that $\sup_\varepsilon \|f^\varepsilon\| \leq \mathbf{c}$, we can extract a subsequence $f^{\varepsilon'}$ satisfying $\|\mathcal{A}^{\varepsilon'} f^{\varepsilon'} - w^0\|_{\varepsilon'} \to 0$, as $\varepsilon' \to 0$, for a certain $w^0 \in \mathcal{W}$.*

Let $\{\mu_i^\varepsilon\}_{i=1}^\infty$ ($\{\mu_i^0\}_{i=1}^\infty$, respectively) be the sequence of the eigenvalues of $\mathcal{A}^\varepsilon$ ($\mathcal{A}^0$, respectively) with the usual convention of repeated eigenvalues. Let $\{w_i^\varepsilon\}_{i=1}^\infty$ ($\{w_i^0\}_{i=1}^\infty$, respectively) be the corresponding eigenfunctions that are assumed to form an orthonormal basis in $\mathbf{H}$.

Then, for each fixed k, we have that $\mu_k^\varepsilon \to \mu_k^0$, as $\varepsilon \to 0$. In addition, for each sequence, still denoted by ε, we can extract a subsequence $\varepsilon' \to 0$ such that

$$\|\mathcal{A}^{\varepsilon'} w_k^{\varepsilon'} - w_k^*\| \to 0 \quad as \quad \varepsilon' \to 0,$$

where w_k^* is an eigenfunction of $\mathcal{A}^0$ associated with μ_k^0, and the set $\{w_i^*\}_{i=1}^\infty$ forms an orthogonal basis of $\mathbf{H}$.

Theorem 3 *Let $r_0 > 0$ and $\beta^0 > 0$ in (2) and (3), respectively. Then, for each fixed k, $k = 1, 2, 3 \ldots$, λ_k^ε in (11) and λ_k^0 in (30) satisfy*

$$\lambda_k^\varepsilon \to \lambda_k^0, \;\; as\, \varepsilon \to 0,$$

where $\{\lambda_k^0\}_{k=1}^\infty$ are the eigenvalues of (29) when $M(\hat{x}) = r_0 C^e(\hat{x})$. In addition, for each sequence, we can extract a subsequence, still denoted by ε, such that *the corresponding eigenfunctions u_k^ε converge toward u_k^0 in $L^2(\Omega)$, where u_k^0 is an eigenfunction of (29) associated with λ_k^0, and the set $\{u_k^0\}_{k=1}^\infty$ forms an orthogonal basis of $L^2(\Omega)$.*

Proof Let us introduce the operators $\mathcal{A}^\varepsilon$, $\mathcal{A}^0 : L^2(\Omega) \to L^2(\Omega)$. We set $\mathcal{A}^\varepsilon f = u^\varepsilon$, where $u^\varepsilon \in \mathbf{V}$ is the unique solution of (7). Similarly, we set $\mathcal{A}^0 f = u^0$, where $u^0 \in \mathbf{V}$ is the unique solution of (28). In this way, the eigenelements of $\mathcal{A}^\varepsilon$ are $\{((\lambda_k^\varepsilon)^{-1}, u_k^\varepsilon)\}_{k=1}^\infty$ with $\{(\lambda_k^\varepsilon, u_k^\varepsilon)\}_{k=1}^\infty$ the eigenelements of (10), and the eigenelements of $\mathcal{A}^0$ are $\{((\lambda_k^0)^{-1}, u_k^0)\}_{k=1}^\infty$ with $\{(\lambda_k^0, u_k^0)\}_{k=1}^\infty$ the eigenelements of (29).

We define $\mathcal{W} = \mathbf{V}$, and considering Theorem 2, properties (i1) and (i2) in Lemma 2 became self-evident. To prove property (i3), we consider $f^\varepsilon \in L^2(\Omega)$, bounded in $L^2(\Omega)$ independently of ε, and consequently, there is a subsequence $\varepsilon' \to 0$ and a certain $f \in L^2(\Omega)$ such that $f^{\varepsilon'} \to f^0$ in $L^2(\Omega)$-weak. We replace f by $f^{\varepsilon'}$ in (7), and since (8) also holds, we rewrite the proof of Theorem 2 with minor modifications, to show the convergence of solutions $u^{\varepsilon'}$ toward u^0 in $H^1(\Omega)$-weak, as $\varepsilon' \to 0$, and property (i3) of Lemma 2 also holds.

Consequently, the convergence of the eigenvalues and the associated eigenfunctions in the statement of the theorem holds from Lemma 2.   □

Acknowledgments This work has partially been supported by MICINN PGC2018-098178-B-I00.

References

1. Ansini, N.: The nonlinear sieve problem and applications to thin films. Asymptotic Anal. **39**(2), 113–145 (2004)
2. Attouch, H.: Variational Convergence for Functions and Operators. Applicable Mathematics Series. Pitman, London (1984)
3. Brillard, A., Lobo, M., Pérez, E.: Homogénéisation de Frontières par epi-convergence en élasticité linéare. RAIRO Modél. Math. Anal. Numér. **24**, 5–26 (1990)
4. Chechkin, G.A., Gadyl'shin, R.R.: On boundary-value problems for the Laplacian in bounded domains with micro inhomogeneous structure of the boundaries. Acta Math. Sin. (Engl. Ser.) **23**(2), 237–248 (2007)
5. Cioranescu, D., Damlamian, D., Griso, G., Onofrei, D.: The periodic unfolding method for perforated domains and Neumann sieve models. J. Math. Pures Appl. **89**, 248–277 (2008)
6. Cioranescu, D., Murat, F.: A strange term coming from nowhere. In: Topics in the Mathematical Modelling of Composite Materials. Progr. Nonlinear Differential Equations Appl., vol. 31. Birkäuser, Boston (1997), pp. 45–93
7. Del Vecchio, T.: The thick Neumann's sieve. Ann. Mat. Pura Appl. **147**, 363–402 (1987)
8. Gómez, D., Pérez, E., Shaposhnikova, T.A.: On homogenization of nonlinear Robin type boundary conditions for cavities along manifolds and associated spectral problems. Asymptot. Anal. **80**, 289–322 (2012)
9. Gómez, D., Pérez, E., Shaposhnikova, T.A.: On correctors for spectral problems in the homogenization of Robin boundary conditions with very large parameters. Int. J. Appl. Math. **26**, 309–320 (2013)

10. Gómez, D., Pérez, E., Shaposhnikova, T.A.: Spectral boundary homogenization problems in perforated domains with Robin boundary conditions and large parameters. In: Integral Methods in Science and Engineering, pp. 155–174. Birkhäuser/Springer, New York (2013)

11. Gómez, D., Lobo, M., Pérez, E., Sanchez-Palencia, E.: Homogenization in perforated domains: a Stokes grill and an adsorption process. Appl. Anal. **97**, 2893–2919 (2018)

12. Gómez, D., Lobo, M., Pérez-Martínez, M.-E.: Asymptotics for models of non-stationary diffusion in domains with a surface distribution of obstacles. Math. Methods Appl. Sci. **42**, 403–413 (2019)

13. Gómez, D., Pérez, E., Podolskiy, A.V., Shaposhnikova, T.A.: Homogenization of variational inequalities for the p-Laplace operator in perforated media along manifolds. Appl. Math. Optim. **79**, 695–713 (2019)

14. Gómez, D., Nazarov, S.A., Pérez-Martínez, M.-E.: Spectral homogenization problems in linear elasticity with large reaction terms concentrated in small regions of the boundary: In: Computational and Analytic Methods in Science and Engineering, p. 127–150 Birkäuser/Springer, New York (2020)

15. Gustafson, K., Abe, T.: The third boundary condition - was it Robin's?. Math. Intell. **20**(1), 63–71 (1998)

16. Landau, L., Lifchitz, E.: Physique Théorique. Tome 7. Théorie de l'Élasticité. Mir, Moscow (1990)

17. Leguillon, D., Sanchez-Palencia, E.: Computation of Singular Solutions in Elliptic Problems and Elasticity. Masson, Paris (1987)

18. Lobo, M., Oleinik, O.A., Pérez, M.E., Shaposhnikova, T.A.: On homogenization of solutions of boundary value problems in domains, perforated along manifolds. Ann. Scuola Norm. Sup. Pisa Cl. Sci. 4e série **25**, 611–629 (1997)

19. Lobo, M., Pérez, E.: Asymptotic behaviour of an elastic body with a surface having small stuck regions. RAIRO Modél. Math. Anal. Numér. **22**, 609–624 (1988)

20. Lobo, M., Pérez, E.: On the vibrations of a body with many concentrated masses near the boundary. Math. Models Methods Appl. **3**(2), 249–273 (1993)

21. Lobo, M., Pérez, E.: The skin effect in vibrating systems with many concentrated masses. Math. Methods Appl. Sci. **24**, 59–80 (2001)

22. Marchenko, V.A., Khruslov, E.Ya.: Homogenization of Partial Differential Equations. Birkhäuser, Boston, MA (2006)

23. Murat, F.: The Neumann sieve. In: Nonlinear Variational Problems (Isola d'Elba, 1983). Research Notes in Mathematics, vol. 127, pp. 24–32. Pitman, Boston, MA (1985)

24. Oleinik, O.A., Shamaev, A.S., Yosifian, G.A.: Mathematical Problems in Elasticity and Homogenization. North-Holland, London (1992)

25. Pérez, E., Shaposhnikova, T.A.: Dokl. Math. **85**(2), 198–203 (2012)

26. Pérez-Martínez, M.-E.: Problemas de homogeneización de fronteras en elasticidad lineal. PhD Thesis, Universidad de Cantabria, Santander (1987)

27. Picard, C.: Analyse limite d'équations variationnelles dans un domaine contenant une grille. RAIRO Modél. Math. Anal. Numér. **21**, 293–326 (1987)

28. Sanchez-Hubert, J., Sanchez-Palencia, E.: Vibration and Coupling of Continuous Systems. Asymptotic Methods. Springer, Heidelberg (1989)

29. Sanchez-Palencia, E.: Boundary value problems in domains containing perforated walls. In: Nonlinear Partial Differential Equations and their Applications. Collège de France Seminar, Vol. III. Research Notes in Mathematics, vol. 70. Pitman, Boston (1982), pp. 309–325

30. Trèves, F.: Basic Linear Partial Differential Equations. Academic Press, New York (1975)

Quasilinear Elliptic Problems in a Two-Component Domain with L^1 Data

Rheadel G. Fulgencio and Olivier Guibé

Abstract In this chapter, we consider the following class of quasilinear equations:

$$\begin{cases} -\operatorname{div}(B(x, u_1)\nabla u_1) = f & \text{in } \Omega_1, \\ -\operatorname{div}(B(x, u_2)\nabla u_2) = f & \text{in } \Omega_2 \\ u_1 = 0 & \text{on } \partial\Omega, \\ (B(x, u_1)\nabla u_1)\nu_1 = (B(x, u_2)\nabla u_2)\nu_1 & \text{on } \Gamma, \\ (B(x, u_1)\nabla u_1)\nu_1 = -h(x)(u_1 - u_2) & \text{on } \Gamma. \end{cases}$$

The domain Ω is composed of two components, Ω_1 and Ω_2, with Γ denoting the interface between them. The given function f belongs to $L^1(\Omega)$. We first present a definition of a renormalized solution for this class of equations. The main result of this chapter is the existence of such a solution.

1 Introduction

In this chapter, we study the existence of a solution $u := (u_1, u_2)$ of the following class of quasilinear equations:

R. G. Fulgencio
University of the Philippines - Diliman, Quezon City, Philippines

R. G. Fulgencio
University of Rouen Normandie, Rouen, France
e-mail: rheadel.fulgencio@univ-rouen.fr; rfulgencio@math.upd.edu.ph

O. Guibé (✉)
University of Rouen Normandie, Rouen, France
e-mail: olivier.guibe@univ-rouen.fr

© The Author(s), under exclusive license to Springer Nature Switzerland AG 2021
P. Donato, M. Luna-Laynez (eds.), *Emerging Problems in the Homogenization
of Partial Differential Equations*, SEMA SIMAI Springer Series 10,
https://doi.org/10.1007/978-3-030-62030-1_4

59

$$\begin{cases} -\operatorname{div}(B(x, u_1)\nabla u_1) = f & \text{in } \Omega_1, \\ -\operatorname{div}(B(x, u_2)\nabla u_2) = f & \text{in } \Omega_2 \\ u_1 = 0 & \text{on } \partial\Omega, \\ (B(x, u_1)\nabla u_1)\nu_1 = (B(x, u_2)\nabla u_2)\nu_1 & \text{on } \Gamma, \\ (B(x, u_1)\nabla u_1)\nu_1 = -h(x)(u_1 - u_2) & \text{on } \Gamma. \end{cases} \qquad (P)$$

Here, Ω is our two-component domain and $\partial\Omega$ is its boundary. The open sets Ω_1 and Ω_2 are the two disjoint components of Ω, Γ is the interface between them (see Figure 1), and the vector ν_i is the unit outward normal to Ω_i. The matrix field $B(x, r)$ is coercive and not restricted by any growth condition with respect to r ($B(x, r)$ is bounded on any compact set of $\mathbb{R}$), and the data f is an L^1-function. On the boundary $\partial\Omega$, we have a Dirichlet boundary condition, while on the interface Γ, we have a continuous flux, and the jump of the solution is proportional to the flux. We refer to [7] for a justification of the model in the case of the conduction of heat in solids.

The existence and uniqueness of solution of (P) when $f \in L^2(\Omega)$ was studied in [3, 11, 13]. In [11, 13], the equations are linear, that is, the matrix field B does not depend on the solution u, while in [3], the equations are quasilinear, which is also the case in this study. The above mentioned papers are all motivated by homogenization, which is also our main goal (see [14]).

Since we consider in this chapter an L^1 data, we need an appropriate notion of solution. Let us recall that, for the elliptic equation

$$-\operatorname{div}(A(x, u)\nabla u) = f$$

with Dirichlet boundary condition, if the matrix A is bounded, a solution in the sense of distribution exists (see [6]), but it is not unique in general (see the counterexamples in [22, 23]). If the matrix field is not bounded, then we cannot

Fig. 1 The two-component domain Ω

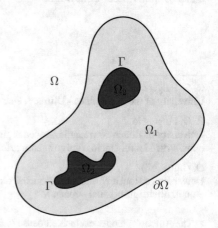

expect to have a solution in the sense of distribution since there is no reason to have $A(x, u) \in L^1_{loc}$. In this chapter, we use the notion of renormalized solution, which was first discussed in [10] by R.J. DiPerna and P.L. Lions for first-order equations. This notion was then further developed by F. Murat in [20], P.L. Lions and F. Murat in [18] for elliptic equations with Dirichlet boundary conditions and L^1 data, and G. Dal Maso et al. in [8] for elliptic equations with general measure data. There is a wide range of literature for elliptic equations with Dirichlet boundary condition and L^1 data, among them are [4, 6, 8, 9, 18, 20]. Considering elliptic equations with Neumann or Robin boundary conditions and L^1 data, which are connected to our problem, gives, in general, additional difficulties due to the lack of the Poincaré inequality or the low regularity of the solution (definition of the trace for example). In the case of one-component domain, L^1 data, and Neumann or Robin boundary conditions, let us mention [1, 2, 21] using the framework of entropy solutions, [15] using a duality method, and [5, 17] using the framework of renormalized solutions.

The main originality of this chapter is the jump of the solution which produces in the formulation a term in the interface Γ. Recalling that the regularity of the renormalized solution is given through the truncate, the first difficulty is to give a sense on the interface for functions (u_1, u_2) whose truncates belong to H^1 only in each component. Following the ideas of [1, 17] (but in the case of one-component domain), we define an appropriate notion of trace (see Proposition 2). The second difficulty is the regularity of $\gamma_1(u_1) - \gamma_2(u_2)$ (where γ_1 is the trace function for $H^1(\Omega_1)$-functions and γ_2 is the trace function for $H^1(\Omega_2)$-functions), since we have to deal, in the renormalized formulation, with terms on the boundary like $(\gamma_1(u_1) - \gamma_2(u_2))S(\gamma(u_1))$, where S is a C^1 function with compact support. To have $(\gamma_1(u_1) - \gamma_2(u_2))S(\gamma(u_1))$ belonging to $L^1(\Gamma)$ is then equivalent to have $\gamma_2(u_2)S(\gamma(u_1)) \in L^1(\Gamma)$, which is unusual and in some sense a coupled regularity on the boundary. It is worth noting that it is not a direct consequence of $T_k(u_1) \in H^1(\Omega_1)$ and $T_k(u_2) \in H^1(\Omega_2)$ (T_k is the usual truncation function at height $\pm k$, see (3)). Using the structure of the equation, we impose an extra regularity (see (11b)), namely

$$(\gamma_1(u_1) - \gamma_2(u_2))[T_k(\gamma(u_1)) - T_k(\gamma(u_2))] \in L^1(\Gamma), \quad \text{for any } k > 0,$$

which allows one to prove that $\gamma_2(u_2)S(\gamma(u_1)) \in L^1(\Gamma)$ and then $(\gamma_1(u_1) - \gamma_2(u_2))S(\gamma(u_1)) \in L^1(\Gamma)$ (see Remark 2). We also impose a decay of the energy of the trace (see (12b)) in addition to the usual decay of the energy, which is crucial to obtain stability results (see Remark 6). Consequently, we are able to give a definition of renormalized solution for problem (P) for which we prove the existence (see Theorem 1).

This chapter is organized as follows. The next section discusses the assumptions on our problem and some definitions, including the definition of a renormalized solution of (P) (see Definition 1). Section 3 is devoted to the proof of the existence of a renormalized solution for (P). We also remark here that by using the Boccardo–Gallouët estimates, we can actually replace conditions (11b) and (12b) of Definition 1 by another regularity condition on the interface. However, we prefer not to use these estimates because we have the homogenization process in mind (see Remark 5).

2   Assumptions and Definitions

In this section, we present the assumptions and definitions necessary for our problem. We begin by introducing the two-component domain Ω. The domain Ω is a connected bounded open set in $\mathbb{R}^N$ with boundary $\partial\Omega$. We can write Ω as the disjoint union $\Omega = \Omega_1 \cup \Omega_2 \cup \Gamma$, where Ω_2 is an open set such that $\overline{\Omega_2} \subset \Omega$ with a Lipschitz boundary Γ, and $\Omega_1 = \Omega \setminus \overline{\Omega_2}$. We denote by ν_i the unit outward normal to Ω_i.

If we have a function u defined on $\Omega \setminus \Gamma$, then we denote by $u_i = u\big|_{\Omega_i}$ the restriction of u in Ω_i. Furthermore, we have the following assumptions:

(A1)   The data f belongs to $L^1(\Omega)$.

(A2)   The function h satisfies

$$h \in L^\infty(\Gamma) \quad \text{and} \quad 0 < h_0 < h(y) \text{ a.e. on } \Gamma, \tag{1}$$

for some $h_0 \in \mathbb{R}^+$.

(A3)   The matrix field B is a Carathéodory function, that is,

    a. the map $r \mapsto B(x, r)$ is continuous for a.e. $x \in \Omega$;
    b. the map $x \mapsto B(x, r)$ is measurable for a.e. $r \in \mathbb{R}$,

and it has the following properties:

(A3.1) $B(x, r)\xi \cdot \xi \geq \alpha|\xi|^2$, for some $\alpha > 0$, for a.e. $x \in \Omega, \forall r \in \mathbb{R}, \forall \xi \in \mathbb{R}^N$;
(A3.2)   for any $k > 0$, $B(x, r) \in L^\infty(\Omega \times (-k, k))^{N \times N}$.

The space for this class of equations is not a usual L^p-space or a Sobolev space due to the jump on the interface. We need the normed space V defined as follows. Let V_1 be the space defined by

$$V_1 = \{v \in H^1(\Omega_1) : v = 0 \text{ on } \partial\Omega\} \quad \text{with} \quad \|v\|_{V_1} := \|\nabla v\|_{L^2(\Omega_1)}.$$

Define $V := \{v \equiv (v_1, v_2) : v_1 \in V_1 \text{ and } v_2 \in H^1(\Omega_2)\}$, equipped with the norm

$$\|v\|_V^2 := \|\nabla v_1\|_{L^2(\Omega_1)}^2 + \|\nabla v_2\|_{L^2(\Omega_2)}^2 + \|v_1 - v_2\|_{L^2(\Gamma)}^2. \tag{2}$$

Identifying $\nabla v := \widetilde{\nabla v_1} + \widetilde{\nabla v_2}$, we have that $\|v\|_V^2 = \|\nabla v\|_{L^2(\Omega \setminus \Gamma)}^2 + \|v_1 - v_2\|_{L^2(\Gamma)}^2$.

Proposition 1 ([19]) *The norm given in (2) is equivalent to the norm of $V_1 \times H^1(\Omega_2)$, that is, there exist two positive constants c_1 and c_2 such that*

$$c_1 \|v\|_V \leq \|v\|_{V_1 \times H^1(\Omega_2)} \leq c_2 \|v\|_V, \quad \forall v \in V.$$

We now define the function T_k, which is known as the truncation function at height $\pm k$. The function $T_k : \mathbb{R} \longrightarrow \mathbb{R}$ is given by

$$T_k(t) = \begin{cases} -k, & \text{if } t \leq k, \\ t, & \text{if } -k \leq t \leq k, \\ k, & \text{if } t \geq k. \end{cases} \tag{3}$$

This function will be crucial in the definition of a renormalized solution of (P).

In the case of L^1 data, we cannot expect to have the solution u belonging to V. In general, in the framework of renormalized solution, the regularity of the solution is given through the regularity of any truncate. So, it is necessary in our case to define the gradient and the trace of the solution u. For the gradient, we follow the definition given in [4]. For the trace, we have to precise the trace of u_1 on Γ and the one of u_2 on Γ. With respect to [1, 17], we have the additional difficulty for u_2 since we do not have the Poincaré inequality.

Proposition 2 *Let $u = (u_1, u_2) : \Omega \setminus \Gamma \longrightarrow \mathbb{R}$ be a measurable function such that $T_k(u) \in V$ for every $k > 0$.*

1. For $i = 1, 2$, there exists a unique measurable function $G_i : \Omega_i \longrightarrow \mathbb{R}^N$ such that for all $k > 0$,

$$\nabla T_k(u_i) = G_i \chi_{\{|u_i|<k\}} \quad a.e. \text{ in } \Omega_i, \tag{4}$$

where $\chi_{\{|u_i|<k\}}$ denotes the characteristic function of

$$\{x \in \Omega_i : |u_i(x)| < k\}.$$

We define G_i as the gradient of u_i and write $G_i = \nabla u_i$.
2. If

$$\sup_{k \geq 1} \frac{1}{k} \|T_k(u)\|_V^2 < \infty, \tag{5}$$

then there exists a unique measurable function

$$w_i : \Gamma \longrightarrow \mathbb{R}, \quad \text{for } i = 1, 2,$$

such that for all $k > 0$,

$$\gamma_i(T_k(u_i)) = T_k(w_i) \quad a.e. \ in \ \Gamma, \tag{6}$$

where $\gamma_i : H^1(\Omega_i) \longrightarrow L^2(\Gamma)$ is the trace operator. We define the function w_i as the trace of u_i on Γ and set

$$\gamma_i(u_i) = w_i, \quad i = 1, 2.$$

Proof

1. This is proved in [4] (see Lemma 2.1).
2. The case $i = 1$, or more generally, the truncates have a zero trace on a part of the boundary (which allows one to use Poincaré-kind inequality), is presented in [17]. We just have to prove the result for $i = 2$.

   The uniqueness is in the almost everywhere sense. Note that if we find two functions that satisfy (6), then the uniqueness of w_2 is assured by the monotonicity of T_k and the fact that w_2 is finite a.e. in Γ.

   By Proposition 1, we know that

$$\|T_k(u_2)\|_{H^1(\Omega_2)} \leq c_1 \|T_k(u)\|_V,$$

for some positive constant c_1, independent of k. It follows from (5) that

$$\|T_k(u_2)\|^2_{H^1(\Omega_2)} \leq Mk, \tag{7}$$

with $M \in \mathbb{R}^+$ independent of k. Due to the regularity of Γ, $\gamma_2(T_n(u_2))$ is well defined and

$$k^2 \mathrm{meas}_\Gamma\{|\gamma_2(T_k(u_2))| \geq k\} = \int_{\Gamma \cap \{|T_k(u_2)| \geq k\}} (\gamma_2(T_k(u_2)))^2 \, d\sigma$$

$$\leq \|\gamma_2(T_k(u_2))\|^2_{L^2(\Gamma)}.$$

Hence, by Trace Theorem and (7), we have

$$k^2 \mathrm{meas}_\Gamma\{|\gamma_2(T_k(u_2))| \geq k\} \leq \|\gamma_2(T_k(u_2))\|^2_{L^2(\Gamma)}$$

$$\leq \|T_k(u_2)\|^2_{L^2(\Omega_2)} + \|\nabla T_k(u_2)\|^2_{L^2(\Omega_2)}$$

$$\leq Mk.$$

As a result,

$$\text{meas}_\Gamma\{|\gamma_2(T_k(u_2))| \geq k\} \longrightarrow 0 \text{ as } k \longrightarrow 0. \tag{8}$$

Define $\Gamma_n = \{x \in \Gamma : |\gamma_2(T_n(u_2))| < n\}$ for $n \in \mathbb{N}$.
From (8), it follows that

$$\Gamma = \bigcup_{n \geq 1} \Gamma_n \cup A, \tag{9}$$

where A is a subset of Γ with zero measure.

Note that for $k < n$, we have $T_k(T_n(u_2)) = T_k(u_2)$. Fix $k > 0$. Then, for every $n \in \mathbb{N}$ such that $n > k$, we have the following equality:

$$T_k(\gamma_2(T_n(u_2))) = \gamma_2(T_k(T_n(u_2))) = \gamma_2(T_k(u_2)) \quad \text{a.e. on } \Gamma,$$

and then

$$\gamma_2(T_k(u_2)) = \gamma_2(T_n(u_2)) \quad \text{a.e. on } \Gamma_k. \tag{10}$$

Since for every $n_1 \leq n$, we have $\Gamma_{n_1} \subseteq \Gamma_n$, in view of (9) and (10), we can define w_2 in the following way:

$$w_2 = \gamma_2(T_n(u_2)) \quad \text{on } \Gamma_n,$$

and noting that $\Gamma = \bigcup_{n \geq 1} \Gamma_n$ (up to measure zero set), we have for any $k > 0$

$$\gamma_2(T_k(u_2)) = T_k(w_2) \quad \text{a.e. on } \Gamma.$$

This concludes the proof. □

Remark 1 In the following, we give an example of a measurable function u where $T_k(u) \in V$, but u_2 is not defined on a part of the interface. We consider $\Omega = (-1, 2)$ with $\Omega_1 = (-1, 0) \cup (1, 2)$ and $\Omega_2 = (0, 1)$ (so $\Gamma = \{0, 1\}$), and $u = (u_1, u_2)$ is defined as

$$u(x) = \begin{cases} u_1(x) = (x+1)(x-2) & \text{if } x \in \Omega_1 \\ u_2(x) = x^{-2} & \text{if } x \in \Omega_2. \end{cases}$$

We have for some positive constants C_1 and C_2

$$\|\nabla T_k(u_1)\|^2_{L^2(\Omega_1)} = \int_{\{|u_1|<k\}} (2x-1)^2 \, dx \leq \int_{\Omega_1} (2x-1)^2 \, dx \leq C_1,$$

and

$$\|\nabla T_k(u_2)\|_{L^2(\Omega_2)}^2 = \int_{k^{1/2}}^1 (-2x^{-3})^2\,dx = 4\left[-\frac{x^7}{7}\right]_{x=k^{1/2}}^1 = \frac{4}{7}(k^{7/2}-1).$$

Thus, we can see that

$$\frac{k^{7/2}}{C} \leq \|T_k(u)\|_V^2 \leq Ck^{7/2},$$

for some $C > 0$, but clearly u_2 does not have a trace on $\{0\} \subset \Gamma$.

We are now in a position to give the definition of renormalized solution.

Definition 1 Let $u = (u_1, u_2) : \Omega \setminus \Gamma \longrightarrow \mathbb{R}$ be a measurable function. Then, u is a renormalized solution of (P) if

$$T_k(u) \in V, \quad \forall k > 0; \tag{11a}$$

$$(u_1 - u_2)(T_k(u_1) - T_k(u_2)) \in L^1(\Gamma), \quad \forall k > 0; \tag{11b}$$

$$\lim_{n\to\infty} \frac{1}{n} \int_{\{|u|<n\}} B(x, u)\nabla u \cdot \nabla u\,dx = 0; \tag{12a}$$

$$\lim_{n\to\infty} \frac{1}{n} \int_\Gamma (u_1 - u_2)(T_n(u_1) - T_n(u_2))\,d\sigma = 0; \tag{12b}$$

and for any $S_1, S_2 \in C^1(\mathbb{R})$ (or equivalently for any $S_1, S_2 \in W^{1,\infty}(\mathbb{R})$) with compact support, u satisfies

$$\int_{\Omega_1} S_1(u_1)B(x, u_1)\nabla u_1 \cdot \nabla v_1\,dx + \int_{\Omega_1} S_1'(u_1)B(x, u_1)\nabla u_1 \cdot \nabla u_1\,v_1\,dx$$

$$+ \int_{\Omega_2} S_2(u_2)B(x, u_2)\nabla u_2 \cdot \nabla v_2\,dx + \int_{\Omega_2} S_2'(u_2)B(x, u_2)\nabla u_2 \cdot \nabla u_2\,v_2\,dx$$

$$+ \int_\Gamma h(x)(u_1 - u_2)(v_1 S_1(u_1) - v_2 S_2(u_2))\,d\sigma$$

$$= \int_{\Omega_1} f v_1 S_1(u_1)\,dx + \int_{\Omega_2} f v_2 S_2(u_2)\,dx, \tag{13}$$

for all $v \in V \cap (L^\infty(\Omega_1) \times L^\infty(\Omega_2))$.

Remark 2 Conditions (11a) (the regularity of the truncate) and (12a) (the decay of the "truncated energy") are standard in the framework of renormalized solutions. As mentioned in the introduction, the main originality of this chapter is the presence of the traces in conditions (11b) and (12b).

In view of Proposition 2, $\gamma(u_1)$ and $\gamma(u_2)$ are well defined. Condition (11b) is an extra regularity of $(u_1 - u_2)(T_k(u_1) - T_k(u_2))$.

Indeed, $(u_1 - u_2)(T_k(u_1) - T_k(u_2))$ cannot be written as

$$(u_1 - u_2)(T_k(u_1) - T_k(u_2))\chi_{\{|u_1|<n\}}\chi_{\{|u_2|<n\}},$$

for any $n \in \mathbb{N}$, so that having $(u_1 - u_2)(T_k(u_1) - T_k(u_2))$ belonging to $L^1(\Gamma)$ is not a consequence of (11a).

Conditions (11a) and (11b) allow one to give a sense of all the terms in (13). Let $S_i \in C^1(\mathbb{R})$, $i = 1, 2$, with compact support. Then, for all $v \in V \cap (L^\infty(\Omega_1) \times L^\infty(\Omega_2))$, we have if supp $S_i \subset [-k, k]$ $(i = 1, 2)$, then for $i = 1, 2$,

$$S_i(u_i)B(x, u_i)\nabla u_i \cdot \nabla v_i = S_i(u_i)B(x, T_k(u_i))\nabla T_k(u_i) \cdot \nabla v_i \in L^1(\Omega_i),$$

and

$$S_i'(u_i)B(x, u_i)\nabla u_i \cdot \nabla u_i\, v_i = S_i'(u_i)B(x, T_k(u_i))\nabla T_k(u_i) \cdot \nabla T_k(u_i)\, v_i \in L^1(\Omega_i),$$

$$f v_i S_i(u_i) \in L^1(\Omega_i).$$

For the boundary term, for any $n \in \mathbb{N}$, let us define $\theta_n : \mathbb{R} \longrightarrow \mathbb{R}$ by

$$\theta_n(s) = \begin{cases} 0, & \text{if } s \leq -2n, \\ \dfrac{s}{n} + 2, & \text{if } -2n \leq s \leq -n, \\ 1, & \text{if } -n \leq s \leq n, \\ -\dfrac{s}{n} + 2, & \text{if } n \leq s \leq 2n, \\ 0, & \text{if } s \geq 2n. \end{cases} \tag{14}$$

Then, since S_1 has a compact support, for some large enough n, we have

$$h(u_1 - u_2)v_1 S_1(u_1) = hv_1(u_1 - u_2)(S_1(u_1) - S_1(u_2))\theta_n(u_1)$$
$$+ hv_1(u_1 - u_2)S_1(u_2)\theta_n(u_1).$$

Since both S_1 and θ_n have compact support, we have that $hv_1(u_1 - u_2)S_1(u_2)\theta_n(u_1)$ is bounded and therefore in $L^1(\Gamma)$. Moreover, since

$$S_1(u_1) - S_1(u_2) = S_1(T_{2n}(u_1)) - S_1(T_{2n}(u_2))$$

and S_1 is Lipschitz, we have

$$|hv_1(u_1 - u_2)(S_1(u_1) - S_1(u_2))\theta_n(u_1)| \leq \|hv_1\|_{L^\infty(\Gamma)}\|S_1'\|_{L^\infty(\mathbb{R})}$$
$$\times |u_1 - u_2||T_{2n}(u_1) - T_{2n}(u_2)|,$$

a.e. in Γ. Thus, in view of (11b), $h(u_1 - u_2)v_1 S_1(u_1) \in L^1(\Gamma)$. Similarly, $h(u_1 - u_2)v_2 S_2(u_2) \in L^1(\Gamma)$.

It is worth noting that condition (11b) is equivalent to have

$$u_2\chi_{\{|u_1|<k\}} \in L^1(\Gamma) \qquad \text{and} \qquad u_1\chi_{\{|u_2|<k\}} \in L^1(\Gamma), \tag{15}$$

for any $k > 0$. Indeed,

$$u_2\chi_{\{|u_1|<k\}} = (u_2 - u_1)\chi_{\{|u_1|<k\}}(\theta_n(u_1) - \theta_n(u_2)) + u_2\theta_n(u_2)\chi_{\{|u_1|<k\}}$$
$$+ u_1\theta_n(u_1)\chi_{\{|u_1|<k\}} - u_1\theta_n(u_2)\chi_{\{|u_1|<k\}},$$

and by condition (11b), the first term on the right-hand side belongs to $L^1(\Gamma)$, while the next 3 terms are bounded and thus also belong to $L^1(\Gamma)$.

Finally, let us comment that conditions (12a) and (12b) are crucial to recover that formally, for any $k > 0$, $T_k(u)$ is an admissible function in (P), that is,

$$\int_{\Omega_1} B(x, u_1)\nabla u_1 \nabla T_k(u_1)\, dx + \int_{\Omega_2} B(x, u_2)\nabla u_2 \nabla T_k(u_2)\, dx$$
$$+ \int_{\Gamma} h(x)(u_1 - u_2)(T_k(u_1) - T_k(u_2))\, d\sigma = \int_{\Omega} f T_k(u_1)\, dx.$$

To prove this, fix $k > 0$. For $n \in \mathbb{N}$, using $S_1 = S_2 = \theta_n$ and $v = T_k(u)$ as a test function in (13), we have

$$\int_{\Omega_1} \theta_n(u_1)B(x, u_1)\nabla u_1 \cdot \nabla T_k(u_1)\, dx$$

$$+ \int_{\Omega_1} \theta_n'(u_1)B(x, u_1)\nabla u_1 \cdot \nabla u_1\, T_k(u_1)\, dx$$

$$+ \int_{\Omega_2} \theta_n(u_2)B(x, u_2)\nabla u_2 \cdot \nabla T_k(u_2)\, dx$$

$$+ \int_{\Omega_2} \theta_n'(u_2)B(x, u_2)\nabla u_2 \cdot \nabla u_2\, T_k(u_2)\, dx$$

$$+ \int_{\Gamma} h(x)(u_1 - u_2)(\theta_n(u_1)T_k(u_1) - \theta_n(u_2)T_k(u_2))\, d\sigma$$

$$= \int_{\Omega} f T_k(u)\theta_n(u)\, dx. \tag{16}$$

Condition (12a) allows one to pass to the limit of the second and fourth integrals in (16), while condition (12b) is useful for passing to the limit of the integral on the boundary in (16).

Remark 3 As observed in the previous remark, the main purpose of introducing condition (11b) is to allow us to make sense of the integral on the interface. We can avoid introducing this extra regularity condition on Γ by using the Boccardo–Gallouët estimates presented in [6]. However, these estimates are heavily dependent on the Sobolev constants. With the final aim of doing the homogenization process, we try as much as possible to refrain from using these estimates (see Remark 5).

Remark 4 In the variational case (i.e. B is a bounded matrix field and $f \in L^q$ with $q \geq (N + 2)/(2N)$), if $B(x, r)$ is global Lipschitz continuous with respect to r or if its modulus of continuity is strongly controlled, the (variational) solution is unique (see [12]). In the L^1 case, the uniqueness question is addressed in [16]: under assumptions (A1)–(A3) and a local Lipschitz condition on $B(x, r)$ with respect to r, we prove that the renormalized solution is unique.

3 Existence Results

In this section, we present the proof for the existence of a renormalized solution of (P).

Theorem 1 *Suppose assumptions (A1)–(A3) hold. Then, there exists a renormalized solution to (P) in the sense of Definition 1.*

Proof The proof is divided into 4 steps. In Step 1, we consider an approximate problem (see (P_ε) below) in which B is approximated by a bounded function and f^ε is an L^2–data. Using Schauder's Fixed Point Theorem, the existence of at least a variational solution of (P_ε) can be shown. Step 2 is devoted to prove some a priori estimates and then extract a convergent subsequence. In Step 3, we prove that conditions (11a), (11b), (12a), and (12b) are satisfied by the limit. Finally, in Step 4, we pass to the limit, and we show that the constructed function is a renormalized solution.

From this point until the end of the proof, we let $i \in \{1, 2\}$.

Step 1: *Introducing the approximate problem and showing the existence of solution of the approximate problem*

Let $\varepsilon > 0$. Suppose $\{f^\varepsilon\} \subset L^2(\Omega)$ such that

$$f^\varepsilon \longrightarrow f \text{ strongly in } L^1(\Omega),$$

as $\varepsilon \to 0$. Define $B_\varepsilon(x, t) = B(x, T_{1/\varepsilon}(t))$. We now consider the following approximate problem:

$$
\begin{cases}
-\operatorname{div}(B_\varepsilon(x, u_1^\varepsilon)\nabla u_1^\varepsilon) = f^\varepsilon & \text{in } \Omega_1, \\
-\operatorname{div}(B_\varepsilon(x, u_2^\varepsilon)\nabla u_2^\varepsilon) = f^\varepsilon & \text{in } \Omega_2, \\
u_1^\varepsilon = 0 & \text{on } \partial\Omega, \\
(B_\varepsilon(x, u_1^\varepsilon)\nabla u_1^\varepsilon)\nu_1 = (B_\varepsilon(x, u_2^\varepsilon)\nabla u_2^\varepsilon)\nu_1 & \text{on } \Gamma, \\
(B_\varepsilon(x, u_1^\varepsilon)\nabla u_1^\varepsilon)\nu_1 = -h(x)(u_1^\varepsilon - u_2^\varepsilon) & \text{on } \Gamma.
\end{cases}
\tag{P_ε}
$$

The variational formulation of problem (P_ε) is the following:

$$
\begin{cases}
\text{Find } u^\varepsilon \in V \text{ such that } \forall\varphi \in V \\
\displaystyle\int_{\Omega_1} B_\varepsilon(x, u_1^\varepsilon)\nabla u_1^\varepsilon \cdot \nabla\varphi_1 \, dx + \int_{\Omega_2} B_\varepsilon(x, u_2^\varepsilon)\nabla u_2^\varepsilon \cdot \nabla\varphi_2 \, dx \\
\quad + \displaystyle\int_\Gamma h(x)(u_1^\varepsilon - u_2^\varepsilon)(\varphi_1 - \varphi_2) \, d\sigma = \int_\Omega f^\varepsilon \varphi \, dx.
\end{cases}
\tag{17}
$$

Using Proposition 1 and Schauder's Fixed Point Theorem, the proof of the existence of solution for (17) is quite standard (see, e.g., [3]).

Step 2: *Extracting subsequences and examining convergences*

Let $u^\varepsilon = (u_1^\varepsilon, u_2^\varepsilon)$ be a solution to the approximate problem (P_ε). By Stampacchia's theorem, for $k > 0$, $T_k(u^\varepsilon) \in V$ since $u^\varepsilon \in V$.

Using $T_k(u^\varepsilon)$ as a test function in the variational formulation (17), we have

$$
\int_{\Omega_1} B_\varepsilon(x, u_1^\varepsilon)\nabla u_1^\varepsilon \nabla T_k(u_1^\varepsilon) \, dx + \int_{\Omega_2} B_\varepsilon(x, u_2^\varepsilon)\nabla u_2^\varepsilon \nabla T_k(u_2^\varepsilon) \, dx
$$

$$
+ \int_\Gamma h(x)(u_1^\varepsilon - u_2^\varepsilon)(T_k(u_1^\varepsilon) - T_k(u_2^\varepsilon)) \, d\sigma = \int_\Omega f^\varepsilon T_k(u^\varepsilon) \, dx.
\tag{18}
$$

By the definition of T_k, the coercivity of B, and the assumption on h, we have

$$
\int_{\Omega_1} B_\varepsilon(x, u_1^\varepsilon)\nabla u_1^\varepsilon \nabla T_k(u_1^\varepsilon) \, dx + \int_{\Omega_2} B_\varepsilon(x, u_2^\varepsilon)\nabla u_2^\varepsilon \nabla T_k(u_2^\varepsilon) \, dx
$$

$$
+ \int_\Gamma h(x)(u_1^\varepsilon - u_2^\varepsilon)(T_k(u_1^\varepsilon) - T_k(u_2^\varepsilon)) \, d\sigma
$$

$$
\geq \alpha\|\nabla T_k(u_1^\varepsilon)\|_{L^2(\Omega_1)}^2 + \alpha\|\nabla T_k(u_2^\varepsilon)\|_{L^2(\Omega_2)}^2 + h_0\|T_k(u_1^\varepsilon) - T_k(u_2^\varepsilon)\|_{L^2(\Gamma)}^2
$$

$$
\geq C_1\|T_k(u^\varepsilon)\|_V^2,
$$

for some positive constant C_1. On the other hand, by Hölder's inequality,

$$\left| \int_{\Omega} f^{\varepsilon} T_k(u^{\varepsilon}) \, dx \right| = \left| \int_{\Omega_1} f^{\varepsilon} T_k(u_1^{\varepsilon}) \, dx + \int_{\Omega_2} f^{\varepsilon} T_k(u_2^{\varepsilon}) \, dx \right|$$

$$\leq \| f^{\varepsilon} \|_{L^1(\Omega)} k \leq Mk,$$

for some positive constant M, which is independent of ε.

Thus,

$$\| T_k(u^{\varepsilon}) \|_V^2 \leq \frac{Mk}{C_1}, \tag{19}$$

that is, the sequence $\{T_k(u^{\varepsilon})\}$ is bounded in V for every $k > 0$.

By the Rellich theorem, the inclusions $V \hookrightarrow L^2(\Omega_1) \times L^2(\Omega_2)$ and $H^{1/2}(\Gamma) \hookrightarrow L^2(\Gamma)$ are compact. Consequently, since $\{T_k(u^{\varepsilon})\}$ is bounded in V for every $k > 0$ (countable), by a diagonal process, we can extract a subsequence of $\{T_k(u^{\varepsilon})\}$ such that for any $k > 0$ (k being a rational number), there is a $v_k \in V$ such that

$$\begin{cases} T_k(u_i^{\varepsilon'}) \longrightarrow v_{k,i} & \text{strongly in } L^2(\Omega_i), \text{ a.e. in } \Omega_i, \\ T_k(u_i^{\varepsilon'}) \rightharpoonup v_{k,i} & \text{weakly in } V, \\ \gamma_i(T_k(u_i^{\varepsilon'})) \longrightarrow \gamma_i(v_{k,i}) & \text{strongly in } L^2(\Gamma), \text{ a.e. on } \Gamma. \end{cases} \tag{20}$$

Now, we show that $\{u_i^{\varepsilon'}\}$ and $\{\gamma_i(u_i^{\varepsilon'})\}$ are Cauchy sequences in measure. For $u_i^{\varepsilon'}$, we follow the arguments developed in [4]. For $\gamma_i(u_i^{\varepsilon'})$, we have additional difficulties that are overcome by using Proposition 1.

Note that we have

$$\| T_k(u_i^{\varepsilon'}) \|_{L^2(\Omega_i)}^2 = \int_{\{|u_i^{\varepsilon'}| \geq k\}} |T_k(u_i^{\varepsilon'})|^2 \, dx + \int_{\{|u_i^{\varepsilon'}| < k\}} |T_k(u_i^{\varepsilon'})|^2 \, dx$$

$$= \int_{\{|u_i^{\varepsilon'}| \geq k\}} k^2 \, dx + \int_{\{|u_i^{\varepsilon'}| < k\}} |u_i^{\varepsilon'}|^2 \, dx.$$

It follows that by the Poincaré inequality, Proposition 1, and (19), we have

$$k^2 \text{meas}\{|u^{\varepsilon'}| \geq k\} = \int_{\{|u_1^{\varepsilon'}| \geq k\}} k^2 \, dx + \int_{\{|u_2^{\varepsilon'}| \geq k\}} k^2 \, dx$$

$$\leq \| T_k(u_1^{\varepsilon'}) \|_{L^2(\Omega_1)}^2 + \| T_k(u_2^{\varepsilon'}) \|_{L^2(\Omega_2)}^2$$

$$\leq C_3 \| \nabla T_k(u_1^{\varepsilon'}) \|_{L^2(\Omega_1)}^2 + \| T_k(u_2^{\varepsilon'}) \|_{H^1(\Omega_2)}^2$$

$$\leq C_4 \| T_k(u^{\varepsilon'}) \|_V^2 \leq \frac{C_4 Mk}{C_1},$$

for some $C_3, C_4 \in \mathbb{R}^+$. Thus, we can find a positive constant C independent of ε such that

$$\text{meas}\{|u_i^{\varepsilon'}| \geq k\} \leq \frac{C}{k}. \tag{21}$$

For $\gamma_1(u_1^{\varepsilon'})$, observe that by the Poincaré inequality and (19),

$$k^2 \text{meas}_\Gamma \{|\gamma_1(u_1^{\varepsilon'})| \geq k\} = \int_{\{|\gamma_1(u_1^{\varepsilon'})| \geq k\}} k^2 \, d\sigma$$

$$= \int_{\{|\gamma_1(u_1^{\varepsilon'})| \geq k\}} \gamma_1(T_k(u_1^{\varepsilon'}))^2 \, d\sigma$$

$$\leq \|\gamma_1(T_k(u_1^{\varepsilon'}))\|_{L^2(\Gamma)}^2$$

$$\leq C_5 \|\nabla T_k(u_1^{\varepsilon'})\|_{L^2(\Omega_2)}^2 \leq C_6 k.$$

Consequently,

$$\text{meas}_\Gamma \{|\gamma_1(u_1^{\varepsilon'})| \geq k\} \leq \frac{C_6}{k} \longrightarrow 0 \quad \text{as } k \longrightarrow \infty. \tag{22}$$

For $\gamma_2(u_2^{\varepsilon'})$, by the Trace Theorem, Proposition 1, and (19), we have

$$k^2 \text{meas}_\Gamma \{|\gamma_2(u_2^{\varepsilon'})| \geq k\} = \int_{\{|\gamma_2(u_2^{\varepsilon'})| \geq k\}} k^2 \, d\sigma$$

$$= \int_{\{|\gamma_2(u_2^{\varepsilon'})| \geq k\}} \gamma_2(T_k(u_2^{\varepsilon'}))^2 \, d\sigma$$

$$\leq \|\gamma_2(T_k(u_2^{\varepsilon'}))\|_{L^2(\Gamma)}^2$$

$$\leq C_7 \|T_k(u_2^{\varepsilon'})\|_{H^1(\Omega_2)}^2 \leq C_8 k.$$

It follows that

$$\text{meas}_\Gamma \{|\gamma_2(u_2^{\varepsilon'})| \geq k\} \leq \frac{C_8}{k} \longrightarrow 0 \quad \text{as } k \longrightarrow \infty. \tag{23}$$

By (22) and (23), for every $\eta > 0$, there exists k_0 such that for every $k \geq k_0$,

$$\text{meas}_\Gamma \{x \in \Gamma; |\gamma_i(u_i^{\varepsilon'})| \geq k\} < \eta. \tag{24}$$

Let $\omega, \eta > 0$. By (21) and (24), we can find large enough k such that

$$\text{meas}\{|u_i^{\varepsilon'}| \geq k\} \leq \frac{\eta}{3}, \tag{25}$$

$$\text{meas}_\Gamma\{x \in \Gamma; |\gamma_i(u_i^{\varepsilon'})| \geq k\} \leq \frac{\eta}{3}, \tag{26}$$

for every $\varepsilon' > 0$. Note that from (20), we can deduce that the sequences $\{T_k(u_i^\varepsilon)\}$ and $\{\gamma_i(T_k(u_i^\varepsilon))\}$ are Cauchy in measure. Hence, there exists $\varepsilon_0 > 0$ such that

$$\text{meas}\{|T_k(u_i^{\varepsilon'}) - T_k(u_i^{\varepsilon''})| \geq \omega\} < \frac{\eta}{3}, \tag{27}$$

$$\text{meas}_\Gamma\{|\gamma_i(T_k(u_i^{\varepsilon'})) - \gamma_i(T_k(u_i^{\varepsilon''}))| \geq \omega\} < \frac{\eta}{3}, \tag{28}$$

for every $0 < \varepsilon', \varepsilon'' < \varepsilon_0$.

Observe that

$$\{|u_i^{\varepsilon'} - u_i^{\varepsilon''}| \geq \omega\} \subset \{|u_i^{\varepsilon'}| \geq k\} \cup \{|u_i^{\varepsilon''}| \geq k\} \cup \{|T_k(u_i^{\varepsilon'}) - T_k(u_i^{\varepsilon''})| \geq \omega\},$$

and thus,

$$\text{meas}\{|u_i^{\varepsilon'} - u_i^{\varepsilon''}| \geq \omega\} \leq \text{meas}\{|u_i^{\varepsilon'}| \geq k\} + \text{meas}\{|u_i^{\varepsilon''}| \geq k\}$$
$$+ \text{meas}\{|T_k(u_i^{\varepsilon'}) - T_k(u_i^{\varepsilon''})| \geq \omega\}.$$

It follows from (25) and (27) that

$$\text{meas}\{|u_i^{\varepsilon'} - u_i^{\varepsilon''}| \geq \omega\} < \eta,$$

that is, $\{u_i^{\varepsilon'}\}$ is actually Cauchy in measure. Using the inequalities (26) and (28), and similar arguments, it can be shown that $\{\gamma_i(u_i^{\varepsilon'})\}$ is Cauchy in measure.

Consequently, there is a subsequence of $\{u_i^{\varepsilon'}\}$ that is convergent a.e. to some measurable function $u_i : \Omega_i \longrightarrow \overline{\mathbb{R}}$, that is,

$$u_i^{\varepsilon'} \longrightarrow u_i \quad \text{a.e. in } \Omega_i. \tag{29}$$

It follows from (21) that u_i is finite a.e. in Ω_i. This $u := (u_1, u_2)$ is our candidate for a renormalized solution for problem (P).

We now prove that u satisfies the conditions (11). Indeed, by the continuity of T_k, we have

$$T_k(u^{\varepsilon'}) \longrightarrow T_k(u) = v_k \in V \quad \text{a.e. in } \Omega \setminus \Gamma. \tag{30}$$

Moreover, we can deduce that $\{\gamma_i(u_i^{\varepsilon'})\}$ is convergent a.e. on Γ up to a subsequence. Hence, there exists $\omega_i : \Gamma \longrightarrow \overline{\mathbb{R}}$ such that

$$\gamma_i(u_i^{\varepsilon'}) \longrightarrow \omega_i \quad \text{a.e. on } \Gamma, \tag{31}$$

with ω_i finite a.e. on Γ by (24). We now identify w_i and $\gamma_i(u_i)$. Using (19) and (20), we obtain

$$\frac{1}{k}\|T_k(u)\|_V^2 \leq \frac{M}{C_1},$$

for any $k > 0$.

By Proposition 2, $\gamma_i(u_i)$ (the trace in the truncate sense) is well defined. From (20), (30), and (31), we obtain that for any $k > 0$,

$$T_k(\omega_i) = \gamma_i(v_{k,i}) = \gamma_i(T_k(u_i)) = T_k(\gamma_i(u_i)) \quad \text{a.e. on } \Gamma.$$

Then, we have $\omega_i = \gamma_i(u_i)$ a.e. on Γ. By Fatou's lemma, T_k being non-decreasing, we have for all $k > 0$

$$\int_\Gamma (u_1 - u_2)(T_k(u_1) - T_k(u_2)) \, d\sigma \leq \liminf_{\varepsilon' \to 0} \int_\Gamma (u_1^{\varepsilon'} - u_2^{\varepsilon'})(T_k(u_1^{\varepsilon'}) - T_k(u_2^{\varepsilon'})) \, d\sigma$$

$$\leq kM,$$

which is (11b).

From this point, we just denote our sequence by ε. Rewriting all the results we got in terms of ε, we have the following: for all $k > 0$,

$$\begin{cases} u_i^\varepsilon \longrightarrow u_i & \text{a.e. in } \Omega, \\ T_k(u_i^\varepsilon) \longrightarrow T_k(u_i) & \text{strongly in } L^2(\Omega_i), \text{ a.e. in } \Omega_i, \\ \gamma_i(u_i^\varepsilon) \longrightarrow \gamma_i(u_i) & \text{a.e. on } \Gamma, \\ \gamma_i(T_k(u_i^\varepsilon)) \longrightarrow \gamma_i(T_k(u_i)) & \text{strongly in } L^2(\Gamma), \text{ a.e. in } \Gamma. \end{cases} \tag{32}$$

In addition, we have

$$\nabla T_k(u_i^\varepsilon) \rightharpoonup \nabla T_k(u_i) \quad \text{weakly in } (L^2(\Omega_i))^N. \tag{33}$$

Step 3: *Showing conditions (12) of Definition 1.*

From the continuity of B and (32), we have that for any fixed $n > 0$,

$$B(x, T_n(u^\varepsilon)) \longrightarrow B(x, T_n(u)) \quad \text{a.e. in } \Omega \text{ and weakly* in } L^\infty(\Omega). \tag{34}$$

Due to Assumption (A3.1) and the lower semi-continuity of the weak convergence,

$$\frac{1}{n} \int_{\{|u|<n\}} B(x, u) \nabla u \cdot \nabla u \, dx = \frac{1}{n} \int_{\Omega \backslash \Gamma} B(x, T_n(u)) \nabla T_n(u) \cdot \nabla T_n(u) \, dx$$

$$\leq \liminf_{\varepsilon \to 0} \frac{1}{n} \int_{\Omega \backslash \Gamma} B(x, T_n(u^\varepsilon)) \nabla T_n(u^\varepsilon) \nabla T_n(u^\varepsilon) \, dx,$$

and by Fatou's lemma,

$$\frac{1}{n} \int_\Gamma (u_1 - u_2)(T_n(u_1) - T_n(u)) \, d\sigma \leq \liminf_{\varepsilon \to 0} \frac{1}{n} \int_\Gamma (u_1^\varepsilon - u_2^\varepsilon)(T_n(u_1^\varepsilon) - T_n(u_2^\varepsilon)) \, d\sigma.$$

Since

$$\int_{\Omega \backslash \Gamma} B(x, T_n(u^\varepsilon)) \nabla T_n(u^\varepsilon) \nabla T_n(u^\varepsilon) \, dx \text{ and } \int_\Gamma (u_1^\varepsilon - u_2^\varepsilon)(T_n(u_1^\varepsilon) - T_n(u_2^\varepsilon)) \, d\sigma$$

are nonnegative, it is sufficient to show that

$$\lim_{n \to \infty} \limsup_{\varepsilon \to 0} \frac{1}{n} \left(\int_{\Omega \backslash \Gamma} B(x, T_n(u^\varepsilon)) \nabla T_n(u^\varepsilon) \nabla T_n(u^\varepsilon) \, dx \right.$$

$$\left. + \int_\Gamma (u_1^\varepsilon - u_2^\varepsilon)(T_n(u_1^\varepsilon) - T_n(u_2^\varepsilon)) \, d\sigma \right) = 0. \tag{35}$$

We use $\frac{1}{n} T_n(u^\varepsilon)$ as a test function in (17) to obtain

$$\frac{1}{n} \int_{\Omega_1} B_\varepsilon(x, u_1^\varepsilon) \nabla u_1^\varepsilon \nabla T_n(u_1^\varepsilon) \, dx + \frac{1}{n} \int_{\Omega_2} B_\varepsilon(x, u_2^\varepsilon) \nabla u_2^\varepsilon \nabla T_n(u_2^\varepsilon) \, dx$$

$$+ \frac{1}{n} \int_\Gamma h(x)(u_1^\varepsilon - u_2^\varepsilon)(T_n(u_1^\varepsilon) - T_n(u_2^\varepsilon)) \, d\sigma = \frac{1}{n} \int_\Omega f^\varepsilon T_n(u^\varepsilon) \, dx.$$

Consequently, for small enough ε, we have

$$\frac{1}{n} \int_{\Omega_1} B(x, T_n(u_1^\varepsilon)) \nabla u_1^\varepsilon \nabla T_n(u_1^\varepsilon) \, dx + \frac{1}{n} \int_{\Omega_2} B(x, T_n(u_2^\varepsilon)) \nabla u_2^\varepsilon \nabla T_n(u_2^\varepsilon) \, dx$$

$$+ \frac{1}{n} \int_\Gamma (u_1^\varepsilon - u_2^\varepsilon)(T_n(u_1^\varepsilon) - T_n(u_2^\varepsilon)) \, d\sigma = \frac{1}{n} \int_\Omega f^\varepsilon T_n(u^\varepsilon) \, dx.$$

Furthermore, since $T_n(u^\varepsilon)$ converges to $T_n(u)$ weakly* in $L^\infty(\Omega)$ and f^ε converges to f in $L^1(\Omega)$, we have

$$\frac{1}{n} \int_\Omega f^\varepsilon T_n(u^\varepsilon) \, dx \longrightarrow \frac{1}{n} \int_\Omega f T_n(u) \, dx \quad \text{as } \varepsilon \longrightarrow 0.$$

It follows that

$$\limsup_{\varepsilon \to 0} \frac{1}{n} \left(\int_{\Omega \backslash \Gamma} B(x, T_n(u^\varepsilon)) \nabla T_n(u^\varepsilon) \nabla T_n(u^\varepsilon) \, dx \right.$$

$$\left. + \int_\Gamma (u_1^\varepsilon - u_2^\varepsilon)(T_n(u_1^\varepsilon) - T_n(u_2^\varepsilon)) \, d\sigma \right) = \frac{1}{n} \int_\Omega f T_n(u) \, dx.$$

Observe that since u is finite a.e.,

$$\frac{1}{n} T_n(u) \longrightarrow 0 \quad \text{a.e. in } \Omega.$$

In addition, for any $n > 0$, $|T_n(u)| \leq n$ a.e., and thus,

$$\left| \frac{1}{n} f T_n(u) \right| \leq |f| \in L^1(\Omega).$$

By the Lebesgue Dominated Convergence Theorem, we obtain

$$\lim_{n \to \infty} \frac{1}{n} \int_\Omega f T_n(u) \, dx = 0,$$

which gives (35).

Step 4. *Show that u satisfies* (13) *of Definition 1.*

Let $S_1, S_2 \in C^1(\mathbb{R})$ with compact support, and let $k > 0$ such that

$$\text{supp } S_i \subset [-k, k]. \tag{36}$$

We need to show that for any $v \in V \cap (L^\infty(\Omega_1) \times L^\infty(\Omega_2))$, u satisfies (13).

We use the function θ_n defined in (14). Note that

$$\theta_n(u_i^\varepsilon) = \theta_n(T_{2n}(u_i^\varepsilon)) \in H^1(\Omega_i) \cap L^\infty(\Omega_i),$$

and thus, if we define

$$\psi_i = v_i S_i(u_i) \theta_n(u_i^\varepsilon),$$

for $v \in V \cap (L^\infty(\Omega_1) \times L^\infty(\Omega_2))$, we have that

$$\psi = (\psi_1, \psi_2) \in V \cap (L^\infty(\Omega_1) \times L^\infty(\Omega_2)).$$

Using ψ as a test function in (17), we have

$$I_{11} + I_{12} + I_{21} + I_{22} + I_{31} + I_{32} + I_4 = I_{51} + I_{52}, \tag{37}$$

where

$$I_{1i} = \int_{\Omega_i} B_\varepsilon(x, u_i^\varepsilon) \nabla u_i^\varepsilon \cdot \nabla v_i \, S_i(u_i) \theta_n(u_i^\varepsilon) \, dx$$

$$I_{2i} = \int_{\Omega_i} B_\varepsilon(x, u_i^\varepsilon) \nabla u_i^\varepsilon \cdot \nabla u_i \, S_i'(u_i) \theta_n(u_i^\varepsilon) \, dx$$

$$I_{3i} = \int_{\Omega_i} B_\varepsilon(x, u_i^\varepsilon) \nabla u_i^\varepsilon \cdot \nabla u_i^\varepsilon \, S_i(u_i) S_n'(u_i^\varepsilon) \, dx$$

$$I_4 = \int_\Gamma h(x)(u_1^\varepsilon - u_2^\varepsilon)(v_1 S_1(u_1)\theta_n(u_1^\varepsilon) - v_2 S_2(u_2)\theta_n(u_2^\varepsilon)) \, d\sigma$$

$$I_{5i} = \int_{\Omega_i} f v_i \, S_i(u_i)\theta_n(u_i^\varepsilon) \, dx.$$

We look at the behavior of each integral. In particular, we will pass to the limit as $\varepsilon \longrightarrow 0$ and then as $n \longrightarrow \infty$.

Note that for $n \geq k$, since supp $S_i \subset [-k, k]$, we have

$$\theta_n(s)S_i(s) = S_i(s) \quad \text{and} \quad \theta_n(s)S_i'(s) = S_i'(s), \quad \text{for a.e. } s \in \mathbb{R}. \tag{38}$$

We first look at I_{1i}. Observe that if ε is small enough, we have

$$B_\varepsilon(x, u_i^\varepsilon)\nabla u_i^\varepsilon \nabla v_i \, S_i(u_i)\theta_n(u_i^\varepsilon) = B(x, T_{1/\varepsilon}(u_i^\varepsilon))\nabla T_{2n}(u_i^\varepsilon) S_i(u_i)\theta_n(u_i^\varepsilon).$$

Choosing ε small enough, we have

$$B(x, T_{1/\varepsilon}(u_i^\varepsilon))\theta_n(u_i^\varepsilon) = B(x, T_{2n}(u_i^\varepsilon))\theta_n(u_i^\varepsilon) \longrightarrow \theta_n(u_i)B(x, T_n(u_i)),$$

a.e. in Ω_i. Moreover, by the assumptions on B, we have

$$|B(x, T_{1/\varepsilon}(u_i^\varepsilon))\theta_n(u_i^\varepsilon)| \leq \sup_{\Omega_i \times [-2n, 2n]} |B(x, r)|.$$

It follows from the Lebesgue Dominated Convergence Theorem that

$$B(x, T_{1/\varepsilon}(u_i^\varepsilon))\theta_n(u_i^\varepsilon) = B(x, T_{2n}(u_i^\varepsilon))\theta_n(u_i^\varepsilon) \longrightarrow \theta_n(u_i)B(x, T_{2n}(u_i)),$$

a.e. in Ω_i and in $L^\infty(\Omega_i)$ weak-$*$. This and (33) imply as $\varepsilon \longrightarrow 0$,

$$I_{1i} \longrightarrow \int_{\Omega_i} B(x, T_{2n}(u_i))\nabla T_{2n}(u_i)\nabla v_i \, S_i(u_i)\theta_n(u_i) \, dx$$

$$= \int_{\Omega_i} B(x, u_i)\nabla u_i \nabla v_i \, S_i(u_i)\theta_n(u_i) \, dx.$$

By (38), we have

$$\lim_{n\to\infty}\lim_{\varepsilon\to 0} I_{1i} = \int_{\Omega_i} B(x, u_i)\nabla u_i \nabla v_i S_i(u_i)\, dx. \tag{39}$$

We now observe the behavior of I_{2i}. For small enough ε, we have

$$B_\varepsilon(x, u_i^\varepsilon)\nabla u_i^\varepsilon \nabla u_i\, v_i\, S_i'(u_i)\theta_n(u_i^\varepsilon) = B(x, T_{2n}(u_i^\varepsilon))\nabla T_{2n}(u_i^\varepsilon)\nabla u_i\, v_i\, S_i'(u_i)\theta_n(u_i^\varepsilon),$$

a.e. in Ω_i.

Since $\nabla u_i\, v_i S_i'(u_i) = \nabla T_{2n}(u_i) v S_i'(u_i) \in (L^2(\Omega_i))^N$, by (33), we obtain, as $\varepsilon \longrightarrow 0$,

$$I_{2i} \longrightarrow \int_{\Omega_i} B(x, T_{2n}(u_i))\nabla T_{2n}(u_i)\nabla u_i\, v_i\, S_i'(u_i)\theta_n(u_i)\, dx$$

$$= \int_{\Omega_i} B(x, u_i)\nabla u_i \nabla u_i\, v_i\, S_i'(u_i)\theta_n(u_i)\, dx.$$

By (38),

$$\lim_{n\to\infty}\lim_{\varepsilon\to 0} I_{2i} = \int_{\Omega_i} B(x, u_i)\nabla u_i \cdot \nabla u_i\, v_i\, S_i'(u_i)\, dx. \tag{40}$$

For the behavior of I_{3i}, we observe that

$$|\theta_n'(s)| \le \frac{1}{n}, \quad \text{for } |s| \le 2n.$$

Consequently,

$$|I_{3i}| \le \frac{\|v_i\|_{L^\infty(\Omega_i)}\|S_i\|_{L^\infty(\mathbb{R})}}{n} \int_{\{|u_i^\varepsilon|<2n\}} B(x, u_i^\varepsilon)\nabla T_{2n}(u_i^\varepsilon)\nabla T_{2n}(u_i^\varepsilon)\, dx.$$

By (35), we have

$$\lim_{n\to\infty}\limsup_{\varepsilon\to 0} I_{3i} = 0. \tag{41}$$

For I_4, we note that

$$h(x)(u_1^\varepsilon - u_2^\varepsilon)v_i S_i(u_i)\theta_n(u_i^\varepsilon) = h(x)(u_1^\varepsilon - u_2^\varepsilon)v_i S_i(u_i)\theta_n(u_i^\varepsilon)\theta_{2n}(u_i^\varepsilon).$$

Then, we can write I_4 as

$$I_4 = I_{41} + I_{42} + I_{43} - I_{44},$$

where

$$I_{41} = \int_\Gamma h(x)(u_1^\varepsilon - u_2^\varepsilon)v_1 S_1(u_1)\theta_{2n}(u_1^\varepsilon)(\theta_n(u_1^\varepsilon) - \theta_n(u_2^\varepsilon))\, d\sigma,$$

$$I_{42} = \int_\Gamma h(x)(u_1^\varepsilon - u_2^\varepsilon)v_1 S_1(u_1)\theta_n(u_2^\varepsilon)\theta_{2n}(u_1^\varepsilon)\, d\sigma,$$

$$I_{43} = \int_\Gamma h(x)(u_1^\varepsilon - u_2^\varepsilon)v_2 S_2(u_2)\theta_{2n}(u_2^\varepsilon)(\theta_n(u_1^\varepsilon) - \theta_n(u_2^\varepsilon))\, d\sigma,$$

$$I_{44} = \int_\Gamma h(x)(u_1^\varepsilon - u_2^\varepsilon)v_2 S_2(u_2)\theta_n(u_1^\varepsilon)\theta_{2n}(u_2^\varepsilon)\, d\sigma.$$

Observe that θ_n is Lipschitz and $\theta_n(u_i^\varepsilon) = \theta_n(T_{2n}(u_i^\varepsilon))$.
This gives

$$|\theta_n(u_1^\varepsilon) - \theta_n(u_2^\varepsilon)| = |\theta_n(T_{2n}(u_1^\varepsilon)) - \theta_n(T_{2n(u_2^\varepsilon)})|$$

$$\leq \frac{1}{n}|T_{2n}(u_1^\varepsilon) - T_{2n}(u_2^\varepsilon)|.$$

Consequently,

$$|I_{41}| \leq \frac{\|h\|_{L^\infty(\Gamma)}\|v_1\|_{L^\infty(\Gamma)}\|S_1\|_{L^\infty(\mathbb{R})}\|\theta_n\|_{L^\infty(\mathbb{R})}}{n}\int_\Gamma |u_1^\varepsilon - u_2^\varepsilon||T_{2n}(u_1^\varepsilon) - T_{2n}(u_2^\varepsilon)|\, d\sigma,$$

and then by (35), we get

$$\lim_{n\to\infty}\limsup_{\varepsilon\to 0} I_{41} = 0. \tag{42}$$

By similar arguments, it can be shown that

$$\lim_{n\to\infty}\limsup_{\varepsilon\to 0} I_{43} = 0. \tag{43}$$

For I_{42}, we observe that

$$|h(u_1^\varepsilon - u_2^\varepsilon)v_1 S_1(u_1)\theta_{2n}(u_1)\theta_n(u_2^\varepsilon)| \leq M,$$

where the constant M depends only on the L^∞-norms of h, S_1, θ_n, and θ_{2n} and n.
Also,

$$h(x)(u_1^\varepsilon - u_2^\varepsilon)v_1 S_1(u_1)\theta_{2n}(u_1^\varepsilon)\theta_n(u_2^\varepsilon) \longrightarrow h(x)(u_1 - u_2)v_1 S_1(u_1)\theta_{2n}(u_1)\theta_n(u_2),$$

a.e. on Γ as $\varepsilon \longrightarrow 0$. By the Lebesgue Dominated Convergence Theorem, as $\varepsilon \longrightarrow 0$,

$$I_{42} \longrightarrow \int_\Gamma h(x)(u_1 - u_2)v_1 S_1(u_1)\theta_{2n}(u_1)\theta_n(u_2) \, d\sigma,$$

and similarly,

$$I_{44} \longrightarrow \int_\Gamma h(x)(u_1 - u_2)v_2 S_2(u_2)\theta_{2n}(u_2)\theta_n(u_1) \, d\sigma.$$

For large enough n, for $j = 1, 2$, $i \neq j$, we have $S_i(u_i)\theta_{2n}(u_i)\theta_n(u_j) = S_i(u_i)\theta_n(u_j)$. In view of (15) in Remark 2, $(u_1 - u_2)S_i(u_i) \in L^1(\Gamma)$, for $i = 1, 2$, so that by the Lebesgue Dominated Convergence Theorem,

$$\lim_{n\to\infty} \lim_{\varepsilon\to 0} I_{42} = \int_\Gamma h(u_1 - u_2)v_1 S_1(u_1) \, d\sigma, \tag{44}$$

$$\lim_{n\to\infty} \lim_{\varepsilon\to 0} I_{44} = \int_\Gamma h(u_1 - u_2)v_2 S_2(u_2) \, d\sigma. \tag{45}$$

Combining (42)–(45), we conclude that

$$\lim_{n\to\infty} \lim_{\varepsilon\to 0} I_4 = \int_\Gamma h(u_1 - u_2)(v_1 S_1(u_1) - v_2 S_2(u_2)) \, d\sigma. \tag{46}$$

Finally, for I_5, observing that $\theta_n(u_i^\varepsilon)$ weakly converges to $\theta_n(u_i)$ in $L^\infty(\Omega_i)$ weakly$*$ and a.e. in Ω_i, f^ε converges strongly to f in $L^1(\Omega)$, we have

$$I_{5i} = \int_{\Omega_i} f^\varepsilon v_i S_i(u_i)\theta_n(u_i^\varepsilon) \, dx \longrightarrow \int_{\Omega_i} f v_i S_i(u_i)\theta_n(u_i) \, dx.$$

From (38), we have

$$\lim_{n\to\infty} \lim_{\varepsilon\to 0} I_{5i} = \int_{\Omega_i} f v_i S_i(u_i) \, dx. \tag{47}$$

Passing through the limit of (37) and using (39), (40), (41), (46), and (47), we have the desired conclusion.

This concludes the proof for the existence of a renormalized solution. □

Remark 5 As explained in the introduction, it is possible to use the Boccardo–Gallouët estimates, that is, to show that $u^\varepsilon = (u_1^\varepsilon, u_2^\varepsilon)$ is bounded in $W^{1,q}(\Omega_1) \times W^{1,q}(\Omega_2)$ and that leads to $u = (u_1, u_2) \in W^{1,q}(\Omega_1) \times W^{1,q}(\Omega_2)$, for all $q < \dfrac{N}{N-1}$. Such a result may simplify the proof since it implies that $\gamma_1(u_1^\varepsilon), \gamma_2(u_2^\varepsilon) \in$

$W^{1-\frac{1}{q},q}(\Gamma)$, and in particular, $\gamma_1(u_1^\varepsilon)$ and $\gamma_2(u_2^\varepsilon)$ are bounded in $L^{1+\eta}(\Gamma)$, for some small enough η. It follows that we can give another definition of renormalized solution including $u_1 - u_2 \in W^{1-\frac{1}{q},q}(\Gamma)$ instead of (11b), and then (12b) is not necessary since it is a direct consequence of the regularity $u_1 - u_2 \in W^{1-\frac{1}{q},q}(\Gamma)$.

However, since we plan to deal with the periodic homogenization of this problem (see [14]), we cannot use the Boccardo–Gallouët estimates since they are strongly related to the Sobolev constant that may blow up in a varying domain. Moreover, our techniques allow us to consider more general equations (with a nonlinear boundary terms) for which the Boccardo–Gallouët estimates are not useful.

Remark 6 (Stability) By adapting the proof of Theorem 1, it is possible to derive a stability result. More precisely, let us consider u^ε, a renormalized solution of

$$\begin{cases} -\operatorname{div}(B_\varepsilon(x, u_1^\varepsilon)\nabla u_1^\varepsilon) = f^\varepsilon & \text{in } \Omega_1, \\ -\operatorname{div}(B_\varepsilon(x, u_2^\varepsilon)\nabla u_2^\varepsilon) = f^\varepsilon & \text{in } \Omega_2, \\ u_1^\varepsilon = 0 & \text{on } \partial\Omega, \\ (B_\varepsilon(x, u_1^\varepsilon)\nabla u_1^\varepsilon)\nu_1 = (B_\varepsilon(x, u_2^\varepsilon)\nabla u_2^\varepsilon)\nu_1 & \text{on } \Gamma, \\ (B_\varepsilon(x, u_1^\varepsilon)\nabla u_1^\varepsilon)\nu_1 = -h^\varepsilon(x)(u_1^\varepsilon - u_2^\varepsilon) & \text{on } \Gamma, \end{cases} \tag{48}$$

where

1. $f^\varepsilon \in L^1(\Omega)$;
2. $B_\varepsilon(x, t)$ is a Carathéodory matrix verifying

   a. $B_\varepsilon(x, r)\xi \cdot \xi \geq \alpha|\xi|^2$, a.e. $x \in \Omega$, for all $r \in \mathbb{R}$, for any $\xi \in \mathbb{R}^N$, and
   b. for any $k > 0$, $B_\varepsilon(x, r) \in L^\infty(\Omega \times (-k, k))^{N\times N}$;

3. $h^\varepsilon \in L^\infty(\Gamma)$ with $0 < h_0 < h^\varepsilon(y)$ a.e. on Γ and $h^\varepsilon(y) < M$ (uniform), for some $M > 0$.

Let $f \in L^1(\Omega)$, $B : \Omega \times \mathbb{R} \longrightarrow \mathbb{R}^{N\times N}$ is a Carathéodory function, and $h : \Gamma \longrightarrow \mathbb{R}$ with $h \geq 0$. If

$$f^\varepsilon \longrightarrow f \quad \text{strongly in } L^1(\Omega);$$

$$\begin{cases} B_\varepsilon(x, r_\varepsilon) \longrightarrow B(x, r) \\ \text{for every sequence } r_\varepsilon \in \mathbb{R} \text{ such that} \\ r_\varepsilon \longrightarrow r \text{ a.e. on } \mathbb{R}; \end{cases}$$

$$h^\varepsilon \longrightarrow h \quad \text{a.e. in } \Gamma;$$

then, u^ε converges to u a.e., where u is a renormalized solution of

$$
\begin{cases}
-\operatorname{div}(B(x, u_1)\nabla u_1) = f & \text{in } \Omega_1, \\
-\operatorname{div}(B(x, u_2)\nabla u_2) = f & \text{in } \Omega_2 \\
u_1 = 0 & \text{on } \partial\Omega, \\
(B(x, u_1)\nabla u_1)\nu_1 = (B(x, u_2)\nabla u_2)\nu_1 & \text{on } \Gamma, \\
(B(x, u_1)\nabla u_1)\nu_1 = -h(x)(u_1 - u_2) & \text{on } \Gamma.
\end{cases}
$$

The main point is to obtain the a priori estimates of Step 2. In view of (16) in Remark 2, $T_k(u^\varepsilon)$ is an "admissible" test function, so that

$$
\int_{\Omega_1} B_\varepsilon(x, u_1^\varepsilon)\nabla u_1^\varepsilon \cdot \nabla T_k(u_1^\varepsilon)\, dx + \int_{\Omega_2} B_\varepsilon(x, u_2^\varepsilon)\nabla u_2^\varepsilon \cdot \nabla T_k(u_2^\varepsilon)\, dx
$$

$$
+ \int_\Gamma h^\varepsilon(x)(u_1^\varepsilon - u_2^\varepsilon)(T_k(u_1^\varepsilon) - T_k(u_2^\varepsilon))\, d\sigma = \int_\Omega f^\varepsilon T_k(u^\varepsilon)\, dx,
$$

which gives all the necessary estimates. Then, we can extract subsequences so that (32) holds true. In view of conditions on f^ε, B_ε, and h^ε, we can perform Step 3 of the proof of Theorem 1.

Acknowledgments The authors would like to thank the anonymous referees for their insightful comments and helpful suggestions.

References

1. Andreu, F., Mazón, J., Segura de León, S., Toledo, J.: Quasi-linear Elliptic and Parabolic Equations in L^1 with Nonlinear Boundary Conditions. Advances in Mathematical Sciences and Applications, vol. 7, pp. 183–213 (1997)
2. Andreu, F., Mazón, J.M., Segura de León, S., Toledo, J.: Existence and uniqueness for a degenerate parabolic equation with L^1-data.. Trans. Am. Math. Soc. **351**, 285–306 (1999)
3. Beltran, R.: Homogenization of a quasilinear elliptic problem in a two-component domain with an imperfect interface, Master's thesis, University of the Philippines - Diliman, 2014
4. Bénilan, P., Boccardo, L., Gallouët, T., Gariepy, R., Pierre, M., Vazquez, J.L.: An L^1 theory of existence and uniqueness of nonlinear elliptic equations. Ann. Sc. Norm. Sup. Pisa **22**, 240–273 (1995)
5. Betta, M.F., Guibé, O., Mercaldo, A.: Neumann problems for nonlinear elliptic equations with L^1 data. J. Differ. Equations **259**, 898–924 (2015)
6. Boccardo, L., Gallouët, T.: Nonlinear elliptic and parabolic equations involving measure data. J. Funct. Anal. **87**, 149–169 (1989)
7. Carslaw, H.S., Jaeger, J.C.: Conduction of Heat in Solids, vol. VIII, 386 p. Clarendon Press, Oxford (1947)
8. Dal Maso, G., Murat, F., Orsina, L., Prignet, A.: Renormalized solutions of elliptic equations with general measure data. In: Annali della Scuola normale superiore di Pisa, Classe di scienze, vol. 28, pp. 741–808 (1999)

9. Dall'Aglio, A.: Approximated solutions of equations with L^1 data. Application to the h-convergence of quasi-linear parabolic equations. Ann. Mat. Pura Appl. **4**, 170, 207–240 (1996)
10. DiPerna, R.J., Lions, P.L.: On the Cauchy problem for Boltzmann equations: global existence and weak stability. Ann. Math. Sec. Ser. **130**, 321–366 (1989)
11. Donato, P., Hang Le Nguyen, K.: Homogenization of diffusion problems with a nonlinear interfacial resistance. Nonlin. Differential Equations Appl. NoDEA **22**, 1345–1380 (2015)
12. Donato, P., Raimondi, F.: Uniqueness result for a class of singular elliptic problems in two-component domains. J. Elliptic Parabol. Equ. **5**, 349–358 (2019)
13. Donato, P., Le Nguyen, H., Tardieu, R.: The periodic unfolding method for a class of imperfect transmission problems. J. Math. Sci. **176**, 891–927 (2011)
14. Donato, P., Fulgencio, R., Guibé, O.: Homogenization results for quasilinear elliptic problems in a two-component domain with L^1 data (submitted)
15. Droniou, J.: Solving convection-diffusion equations with mixed, Neumann and Fourier boundary conditions and measures as data, by a duality method. Adv. Differ. Equ. **5**, 1341–1396 (2000)
16. Fulgencio, R., Guibé, O.: Uniqueness results for quasilinear elliptic problems in a two-component domain with L^1 data (submitted)
17. Guibé, O., Oropeza, A.: Renormalized solutions of elliptic equations with Robin boundary conditions. Acta Math. Sci. **37**, 889–910 (2017)
18. Lions, P.-L., Murat, F.: Sur les solutions renormalisées d'équations elliptiques (unpublished manuscript)
19. Monsurró, S.: Homogenization of a two-component composite with interfacial thermal barrier. Adv. Math. Sci. Appl. **13**, 43–63 (2003)
20. Murat, F.: Soluciones renormalizadas de EDP elipticas no lineales, Tech. Report R93023, Laboratoire d'Analyse Numérique, Paris VI, 1993.
21. Ouaro, S., Ouedraogo, A.: Entropy solution to an elliptic problem with nonlinear boundary conditions. An. Univ. Craiova, Ser. Mat. Inf. **39**, 148–181 (2012)
22. Prignet, A.: Remarks on existence and uniqueness of solutions of elliptic problems with right hand side measures. Rendiconti Mat. Appl. Ser. VII **15**, 321–337 (1995)
23. Serrin, J.: Pathological solutions of elliptic equations. Ann. Scuola Norm. Sup. Pisa Cl. Sci. **18**, 385–387 (1964)

Homogenization of an Eigenvalue Problem in a Two-Component Domain with Interfacial Jump

Patrizia Donato, Eleanor Gemida, and Editha C. Jose

Abstract This work concerns the asymptotic behaviour of the eigenvalues and eigenvectors of a problem posed on an ε-periodic two-component domain with an imperfect interface. We obtain characterizations of the eigenvalues and give homogenization results using the periodic unfolding method. The eigenvalues of the ε-problem converge to the corresponding eigenvalues of the limit problem, for the whole sequence. The same convergence result is obtained for the corresponding eigenspaces. The convergence for the whole sequence of the corresponding eigenvectors is achieved when the associated homogenized eigenvalue is simple.

1 Introduction

This study deals with the homogenization of a stationary elliptic eigenvalue problem with oscillating coefficients in a domain $\Omega \subset \mathbb{R}^N$, which contains two subdomains Ω_1^ε and Ω_2^ε, separated by an interface Γ^ε. The component Ω_2^ε is the union of the disjoint ε-periodic translated sets εY_2, where $\overline{Y_2}$ is a subset of the reference cell Y. On the other hand, the component Ω_1^ε is connected and defined as $\Omega \backslash \overline{\Omega_2^\varepsilon}$.

The modelled two-component composite presents an interfacial resistance, on which the conormal derivative is assumed to be proportional to the jump of the solution.

P. Donato (✉)
Université de Rouen Normandie, Laboratoire de Mathématiques Raphaël Salem, UMR CNRS 6085, Saint Étienne du Rouvray, France
e-mail: patrizia.donato@univ-rouen.fr

E. Gemida · E. C. Jose
Institute of Mathematical Sciences and Physics, University of the Philippines Los Baños, Laguna, Philippines
e-mail: ebgemida@up.edu.ph; ecjose1@up.edu.ph

© The Author(s), under exclusive license to Springer Nature Switzerland AG 2021
P. Donato, M. Luna-Laynez (eds.), *Emerging Problems in the Homogenization of Partial Differential Equations*, SEMA SIMAI Springer Series 10,
https://doi.org/10.1007/978-3-030-62030-1_5

More precisely, we look for a couple $(u^\varepsilon, \lambda^\varepsilon)$ with $u^\varepsilon \neq 0$ such that

$$\begin{cases} -\mathrm{div}(A^\varepsilon \nabla u^\varepsilon) = \lambda^\varepsilon u^\varepsilon & \text{in } \Omega_1^\varepsilon \cup \Omega_2^\varepsilon, \\ [A^\varepsilon \nabla u^\varepsilon \cdot n_1^\varepsilon] = 0 & \text{on } \Gamma^\varepsilon, \\ A^\varepsilon \nabla u_1^\varepsilon \cdot n_1^\varepsilon = -\varepsilon^\gamma h^\varepsilon [u^\varepsilon] & \text{on } \Gamma^\varepsilon, \\ u_1^\varepsilon = 0 & \text{on } \partial\Omega, \end{cases} \tag{1}$$

where $\gamma < 1$, n_1^ε is the unitary outward normal to Ω_1^ε, $u^\varepsilon = (u_1^\varepsilon, u_2^\varepsilon)$ with $u_i^\varepsilon = u^\varepsilon_{|\Omega_i^\varepsilon}$, $i = 1, 2$, and $[\cdot]$ denotes the jump through the interface Γ^ε. Our aim is first to describe the set of the eigenvalues λ^ε and the corresponding eigenfunctions $u^\varepsilon = (u_1^\varepsilon, u_2^\varepsilon)$ and then to study the asymptotic behaviour of the problem when $\varepsilon \to 0$.

We assume that

$$A^\varepsilon(x) := A\left(\frac{x}{\varepsilon}\right), \qquad h^\varepsilon(x) := h\left(\frac{x}{\varepsilon}\right),$$

where A is a Y-periodic $N \times N$ matrix-valued function, which is bounded, symmetric, and uniformly elliptic. Moreover, we suppose that h is a Y-periodic function in $L^\infty(\Gamma)$ bounded from below by a strictly positive constant.

Let us point out that since the solution is an eigenvector, the function at the right-hand side of the first equation in problem (1) is a function dependent on ε, and not a fixed function in $L^2(\Omega)$.

We first prove that the set of eigenvalues λ^ε in problem (1) is a countable set $\{\lambda_k^\varepsilon\}_{k\in\mathbb{N}}$ in $\mathbb{R}^+$ whose unique accumulation point is $+\infty$, and that there exists a sequence of the corresponding eigenvectors $\{u_n^\varepsilon\}$ in H_γ^ε, which forms a complete orthonormal system in $L^2(\Omega)$.

In the second part of the chapter, by using the periodic unfolding method (see [4–6], and the recent book [7] for a complete presentation), we show that for any $k \in \mathbb{N}$, the eigenvalue λ_k^ε converges to the corresponding one of the homogenized eigenvalue problem, whose eigenvector forms a complete orthonormal system in $L^2(\Omega)$. Also, the eigenspaces of the ε-problem converge to those of the homogenized problem. In particular, for a simple eigenvalue, all the sequence of the eigenfunctions converges.

The results of this chapter are part of the master's thesis of the second author [13]. Let us mention that the case $\gamma = 1$ remains open, due to a lack of compactness of the solution in $L^2(\Omega)$ (see Remark 5 for the details).

The first results on the homogenization of eigenvalue problems were given by L. Boccardo and P. Marcellini [1] in the framework of G-convergence (hence, with no periodicity assumption). Successively, the problem for periodically oscillating coefficients in a fixed domain was completely studied by S. Kesavan [15]. We refer to M. Vanninathan [18] for the corresponding study in periodically perforated domains.

This chapter is organized as follows. In Sect. 2, we consider the elliptic eigenvalue problem in the two-component domain for a fixed ε. We introduce the variational formulation of the problem and the functional setting of the solution space H_γ^ε. We obtain some characterizations of the eigenvalues.

In Sect. 3, we present the periodic unfolding method in a two-component domain with a jump on the interface. Then, we present the homogenization results obtained using such method for an elliptic problem where the data is a sequence of functions $\{f^\varepsilon\}$ converging weakly in $L^2(\Omega)$.

In Sect. 4, we prove some a priori estimates of the eigenvalues and the corresponding eigenvectors. By applying the previous homogenization results to our case, we derive the asymptotic behaviour of the eigenvalue problem.

2 An Elliptic Eigenvalue Problem in a Periodic Two-Component Domain

In this section, we describe an elliptic eigenvalue problem in a two-component domain. More precisely, we consider a problem with oscillating coefficients in a domain $\Omega \subset \mathbb{R}^N$, which is the union of the two periodic subdomains Ω_1^ε and $\overline{\Omega_2^\varepsilon}$, separated by an interface Γ^ε. This periodicity is represented by the parameter ε.

We first introduce the domain and the functional setting of the problem. Then, we describe, for fixed ε, its eigenvalues and the eigenfunctions, by applying some known results on abstract eigenvalue problems. The asymptotic behaviour of the problem, as $\varepsilon \to 0$, is studied in the remaining sections.

2.1 Setting of the Eigenvalue Problem

We construct our two-component domain as follows: let Ω be a non-empty, connected, and bounded open set in $\mathbb{R}^N$, $N \geq 2$, with a Lipschitz continuous boundary $\partial\Omega$, and let $Y = (0, l_1) \times \ldots \times (0, l_N)$, $l_i \in \mathbb{R}$ for all $i = 1, 2, \ldots, N$, be the reference cell.

The sets Y_1 and Y_2 are two non-empty open disjoint subsets of Y such that $\overline{Y_2} \subset Y$ and $Y = Y_1 \cup \overline{Y_2}$. We suppose that the component Y_1 is connected and Y_2 has a Lipschitz continuous boundary Γ with at most a finite number of connected components.

For any $k \in \mathbb{Z}^N$, we set

$$k_l =: (k_1 l_1, \ldots, k_N l_N), \quad Y^k := k_l + Y, \quad Y_i^k := k_l + Y_i, \ i = 1, 2, \quad \Gamma_k := k_l + \Gamma.$$

In other words, Y_i^k and Γ_k are translations of the component Y_i, $i = 1, 2$, and of the boundary Γ, respectively.

We suppose that ε takes its values in a given sequence of positive real numbers that tends to zero.

Fig. 1 The sets Ω_i^ε, $i = 1, 2$

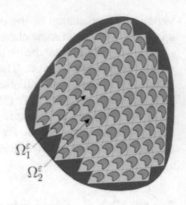

Ω_1^ε

Ω_2^ε

We define the two components as follows:

$$K_\varepsilon = \{k \in \mathbb{Z}^N \mid \varepsilon Y^k \subset \Omega\}, \quad \Omega_2^\varepsilon = \bigcup_{k \in K_\varepsilon} \varepsilon Y_2^k, \quad \Omega_1^\varepsilon = \Omega \backslash \overline{\Omega_2^\varepsilon} \quad i = 1, 2, \quad \Gamma^\varepsilon = \partial \Omega_2^\varepsilon.$$

That is, the disconnected component Ω_2^ε is the union of all the translated sets of εY_2, which belong to the translated cells of εY well contained in Ω, as illustrated in Fig. 1.

We use in the sequel the following notations:

- $\theta_i = |Y_i|/|Y|$, the proportion of material Y_i in Y, $i = 1, 2$,
- $\tilde{u}$, the zero extension to the whole of Ω of a function defined on Ω_1^ε or Ω_2^ε,
- $\mathcal{M}_\omega(f) = \frac{1}{|\omega|} \int_\omega f \, dx$ for any open set $\omega \subset \mathbb{R}^N$ and any $f \in L^1(\omega)$, and
- $M(\alpha, \beta, \Omega)$, the set of $N \times N$ matrix-valued functions in $(L^\infty(\Omega))^{N^2}$ such that

$$A(x)\upsilon\upsilon \geq \alpha|\upsilon|^2 \quad \text{and} \quad |A(x)\upsilon| \leq \beta|\upsilon|, \tag{2}$$

for all $\upsilon \in \mathbb{R}^N$ ($N \geq 2$) and a.e. in Ω, where $\alpha, \beta \in \mathbb{R}, 0 < \alpha < \beta$.

The aim of this work is to study the asymptotic behaviour of the eigenvalue problem:

$$\begin{cases} -\text{div}\,(A^\varepsilon \nabla u_1^\varepsilon) = \lambda^\varepsilon u_1^\varepsilon & \text{in } \Omega_1^\varepsilon, \\ -\text{div}\,(A^\varepsilon \nabla u_2^\varepsilon) = \lambda^\varepsilon u_2^\varepsilon & \text{in } \Omega_2^\varepsilon, \\ A^\varepsilon \nabla u_1^\varepsilon \cdot n_1^\varepsilon = -A^\varepsilon \nabla u_2^\varepsilon \cdot n_2^\varepsilon & \text{on } \Gamma^\varepsilon, \\ -A^\varepsilon \nabla u_1^\varepsilon \cdot n_1^\varepsilon = \varepsilon^\gamma h^\varepsilon (u_1^\varepsilon - u_2^\varepsilon) & \text{on } \Gamma^\varepsilon, \\ u_1^\varepsilon = 0 & \text{on } \partial\Omega, \end{cases} \tag{3}$$

where $\gamma \in \mathbb{R}$ and n_i^ε is the unitary outward normal to Ω_i^ε, $i = 1, 2$.

We suppose that

$$A^\varepsilon(x) := A\left(\frac{x}{\varepsilon}\right) \text{ a.e. in } \Omega, \qquad h^\varepsilon(x) := h\left(\frac{x}{\varepsilon}\right) \text{ a.e. in } \Gamma^\varepsilon, \qquad (4)$$

where

$$\begin{cases} (1) \ A \text{ is in } M(\alpha, \beta, \Omega) \text{ and is symmetric,} \\ (2) \ h \in L^\infty(\Gamma) \text{ and } \exists \, h_0 \in \mathbb{R} \text{ such that } 0 < h_0 < h(y), \quad \text{a.e. on } \Gamma. \end{cases} \qquad (5)$$

Before introducing the variational formulation of problem (3), we discuss the space H_γ^ε and mention some of its properties. We refer to [16] and [9] for the proof of these results.

For any function v defined on Ω, we set

$$v_1 = v_{|\Omega_1^\varepsilon}, \qquad v_2 = v_{|\Omega_2^\varepsilon}. \qquad (6)$$

Hence, for any $v \in L^2(\Omega)$, we have

$$\|v\|_{L^2(\Omega)}^2 = \|v_1\|_{L^2(\Omega_1^\varepsilon)}^2 + \|v_2\|_{L^2(\Omega_2^\varepsilon)}^2,$$

so that we identify $v \in L^2(\Omega)$ with the couple $(v_1, v_2) \in L^2(\Omega_1^\varepsilon) \times L^2(\Omega_2^\varepsilon)$. Conversely, we identify $(v_1, v_2) \in L^2(\Omega_1^\varepsilon) \times L^2(\Omega_2^\varepsilon)$ with $v = \tilde{v}_1 + \tilde{v}_2 \in L^2(\Omega)$.

Accordingly, we have the following definition.

Definition 1 Let V^ε be the space

$$V^\varepsilon = \{u \in H^1(\Omega_1^\varepsilon) : u = 0 \text{ on } \partial\Omega\}$$

together with the norm $\|u\|_{V^\varepsilon} = \|\nabla u\|_{L^2(\Omega_1^\varepsilon)}$.

For every $\gamma \in \mathbb{R}$, we define the functional space H_γ^ε by

$$H_\gamma^\varepsilon = \left\{u \in L^2(\Omega) \mid u = (u_1, u_2) \text{ where } u_1 \in V^\varepsilon, u_2 \in H^1(\Omega_2^\varepsilon)\right\},$$

equipped with the norm

$$\|u\|_{H_\gamma^\varepsilon}^2 = \|\nabla u_1\|_{L^2(\Omega_1^\varepsilon)}^2 + \|\nabla u_2\|_{L^2(\Omega_2^\varepsilon)}^2 + \varepsilon^\gamma \|u_1 - u_2\|_{L^2(\Gamma^\varepsilon)}^2. \qquad (7)$$

Remark 1 In the sequel, we identify the gradient of a function $v \in H_\gamma^\varepsilon$ with its absolutely continuous part. That is, we set

$$\nabla v = \widetilde{\nabla v_1} + \widetilde{\nabla v_2}. \qquad (8)$$

Following (8), we can write the norm in (7) as

$$\|u\|_{H_\gamma^\varepsilon}^2 = \|\nabla u\|_{L^2(\Omega \setminus \Gamma^\varepsilon)}^2 + \varepsilon^\gamma \|u_1 - u_2\|_{L^2(\Gamma^\varepsilon)}^2 .$$

From here on, we denote by C any generic constant independent of ε.

Lemma 1 ([16]) *There exists a constant $C > 0$ independent of ε such that*

$$\|u\|_{L^2(\Omega_1^\varepsilon) \times L^2(\Omega_2^\varepsilon)} \leq C \|u\|_{H_1^\varepsilon} \quad \forall u \in H_1^\varepsilon.$$

Proposition 1 ([9]) *There exists a constant $C > 0$ independent of ε such that*

$$\|u\|_{H_\gamma^\varepsilon}^2 \leq C(1 + \varepsilon^{\gamma-1}) \|u\|_{V^\varepsilon \times H^1(\Omega_2^\varepsilon)}^2, \quad \forall \gamma \in \mathbb{R}, \forall u \in H_\gamma^\varepsilon.$$

If $\gamma \leq 1$, then there exist constants C_1 and C_2 independent of ε such that

$$C_1 \|u\|_{V^\varepsilon \times H^1(\Omega_2^\varepsilon)}^2 \leq \|u\|_{H_\gamma^\varepsilon}^2 \leq C_2(1 + \varepsilon^{\gamma-1}) \|u\|_{V^\varepsilon \times H^1(\Omega_2^\varepsilon)}^2 \quad \text{for all } u \in H_\gamma^\varepsilon.$$

We also have some a priori estimates, such as the next corollary, which are direct consequences of (7).

Corollary 1 ([12]) *Let $u^\varepsilon = (u_1^\varepsilon, u_2^\varepsilon)$ be a sequence such that $\|u^\varepsilon\|_{H_\gamma^\varepsilon}$ is bounded. Then, there exists a constant $C > 0$ independent of ε such that*

(i) $\left\|u_1^\varepsilon\right\|_{H^1(\Omega_1^\varepsilon)} \leq C,$

(ii) $\left\|\nabla u_2^\varepsilon\right\|_{L^2(\Omega_2^\varepsilon)} \leq C,$ *and*

(iii) $\left\|u_1^\varepsilon - u_2^\varepsilon\right\|_{L^2(\Gamma^\varepsilon)} \leq C\varepsilon^{-\frac{\gamma}{2}}.$

Moreover, if $\gamma \leq 1$, then $\left\|u_2^\varepsilon\right\|_{H^1(\Omega_2^\varepsilon)} \leq C.$

Hence, under assumptions (4) and (5), the variational formulation of problem (3) reads

$$\begin{cases} \text{Find } (\lambda^\varepsilon, u^\varepsilon) \in \mathbb{R} \times H_\gamma^\varepsilon \setminus \{0\} \text{ such that} \\[2mm] \displaystyle\int_{\Omega \setminus \Gamma^\varepsilon} A^\varepsilon \nabla u^\varepsilon \nabla v \, dx + \varepsilon^\gamma \int_{\Gamma^\varepsilon} h^\varepsilon (u_1^\varepsilon - u_2^\varepsilon)(v_1 - v_2) \, d\sigma \\[4mm] = \lambda^\varepsilon \displaystyle\int_\Omega u^\varepsilon v \, dx, \quad \text{for all } v \in H_\gamma^\varepsilon. \end{cases} \qquad (9)$$

Remark 2 Note that $\lambda^\varepsilon = 0$ is not an eigenvalue of problem (9) because then, in view of the Lax–Milgram theorem, the only solution of the homogeneous equation is the trivial one, i.e., $u^\varepsilon = 0$, which is not an eigenvector.

2.2 Analysis of the Eigenvalues and Eigenvectors for Fixed ε

In this section, we prove the following result describing, for fixed ε, the eigenvalues and eigenvectors in problem (9):

Theorem 1 *Let $\gamma \in \mathbb{R}$ and ε be fixed. Suppose A^ε and h^ε satisfy assumptions (4) and (5).*
Then,

(i) *the set of eigenvalues λ^ε in problem (9) is a countable set in $\mathbb{R}^+$ whose unique accumulation point is $+\infty$.*
(ii) *Every eigenvalue is of finite multiplicity; that is, its corresponding eigenspace is a vector subspace of $L^2(\Omega)$ of positive finite dimension.*
(iii) *Let $\{\lambda_n^\varepsilon\}$ be the sequence of the eigenvalues in increasing order, where each eigenvalue is repeated as many times as the dimension of its corresponding eigenspace. Then,*

$$0 < \lambda_1^\varepsilon \le \lambda_2^\varepsilon \le \ldots \to +\infty. \tag{10}$$

Moreover, there exists a sequence of the corresponding eigenvectors $\{u_n^\varepsilon\}$ in H_γ^ε, which forms a complete orthonormal system in $L^2(\Omega)$.

Remark 3 The orthonormality of the sequence $\{u_n^\varepsilon\}$ in $L^2(\Omega)$ implies its orthogonality in H_γ^ε under the inner product $\langle \cdot, \cdot \rangle$ defined by

$$\langle \cdot, \cdot \rangle : (u, v) \in H_\gamma^\varepsilon \times H_\gamma^\varepsilon \longmapsto \left(\int_{\Omega \backslash \Gamma^\varepsilon} A^\varepsilon \nabla u \nabla v \, dx + \varepsilon^\gamma \int_{\Gamma^\varepsilon} h^\varepsilon (u_1 - u_2)(v_1 - v_2) \, d\sigma \right) \in \mathbb{R}.$$

Indeed from (9), for every $i, j \in \mathbb{N}$,

$$\left\langle u_i^\varepsilon, u_j^\varepsilon \right\rangle = \int_{\Omega \backslash \Gamma^\varepsilon} A^\varepsilon \nabla u_i^\varepsilon \nabla u_j^\varepsilon \, dx + \varepsilon^\gamma \int_{\Gamma^\varepsilon} h(u_{1i}^\varepsilon - u_{2i}^\varepsilon)(u_{1j}^\varepsilon - u_{2j}^\varepsilon) \, d\sigma$$

$$= \lambda_i^\varepsilon \int_\Omega u_i^\varepsilon u_j^\varepsilon \, dx = \lambda_i^\varepsilon \delta_{ij},$$

where δ_{ij} is the Kronecker delta.

Also, observe that $\|u_i^\varepsilon\|_{H_\gamma^\varepsilon} = \langle u_i^\varepsilon, u_i^\varepsilon \rangle^{1/2} = \left(\lambda_i^\varepsilon \right)^{1/2}$, for every $i \in \mathbb{N}$.

In order to prove Theorem 1, we consider the abstract eigenvalue problem

$$\begin{cases} \text{Find } (\rho, v) \in \mathbb{C} \times H \backslash \{0\} \text{ satisfying} \\ \mathcal{B}v = \rho v, \end{cases} \tag{11}$$

where H is a Hilbert space on $\mathbb{C}$ and $\mathcal{B} : H \longrightarrow H$ such that

$$
\begin{cases}
(i) \ \mathcal{B} \text{ is linear;} \\
(ii) \ \mathcal{B} \text{ is compact, that is, for each bounded sequence } (x_n)_{n \in \mathbb{N}} \text{ in } H, \text{ there} \\
\quad \text{exists a subsequence } (x_{n_k})_{k \in \mathbb{N}} \text{ such that } (\mathcal{B}x_{n_k})_{k \in \mathbb{N}} \text{ is convergent in } H; \\
(iii) \ \mathcal{B} \text{ is symmetric, that is, } \langle \mathcal{B}u, w \rangle_{H \times H} = \langle u, \mathcal{B}w \rangle_{H \times H} \quad \forall u, w \in H; \\
(iv) \ \mathcal{B}u = 0 \text{ implies } u = 0.
\end{cases}
\tag{12}
$$

Let us recall the following well-known result describing the properties of the eigenvalues and eigenvectors of problem (11). We refer to E. Zeidler [19] for a proof.

Theorem 2 *Suppose H is a Hilbert space, and $\mathcal{B} : H \longrightarrow H$, $\mathcal{B} \not\equiv 0$, satisfies all the assumptions in (12). Then,*

(i) *the set of eigenvalues of problem (11) is a countable set in $\mathbb{R}$ whose unique accumulation point is zero.*

(ii) *Every eigenvalue is of finite multiplicity; that is, its corresponding eigenspace is a vector subspace of H of positive finite dimension.*

(iii) *Suppose $\{\rho_n\}$ is the sequence of the eigenvalues in decreasing order, that is, $\rho_n \geq \rho_{n+1} \ \forall \ n \in \mathbb{N}$, where each eigenvalue is repeated as many times as the dimension of its corresponding eigenspace. Then, $\lim_{n \to +\infty} \rho_n = 0$. Moreover, there exists a sequence of the corresponding eigenvectors $\{v_n\}$, which forms a complete orthonormal system in H.*

Proof of Theorem 1 To prove the results, we apply Theorem 2 with $H = L^2(\Omega)$. Since ε is fixed and does not play any role here, we omit it to simplify the notations.

We construct a suitable operator $\mathcal{B}$ as follows. Define the mapping

$$
\Phi : f \in L^2(\Omega) \mapsto u_f \in H_\gamma,
$$

where $u_f = (u_{1f}, u_{2f})$ is the unique solution of the variational problem,

$$
\begin{cases}
\text{Find } u_f \in H_\gamma \text{ such that } \forall \ v \in H_\gamma \\
\int_{\Omega \backslash \Gamma} A \nabla u_f \nabla v \, dx + \int_\Gamma h(u_{1f} - u_{2f})(v_1 - v_2) \, d\sigma = \int_\Omega f v \, dx.
\end{cases}
\tag{13}
$$

The existence and uniqueness of the solution of problem (13) is due to the Lax–Milgram theorem. Moreover, we have the following a priori estimate of the solution:

$$
\| u_f \|_{H_\gamma} \leq C \| f \|_{L^2(\Omega)},
\tag{14}
$$

for some constant $C > 0$.

Now, define the embedding $i : u_f \in H_\gamma \to u_f \in L^2(\Omega)$, which by the Rellich–Kondrachov compactness theorem is compact. Finally, set

$$\mathcal{B} = i \circ \Phi : f \in L^2(\Omega) \mapsto u_f \in L^2(\Omega). \tag{15}$$

Let us check that the mapping $\mathcal{B}$ satisfies the assumptions of Theorem 2.

(*i*) The linearity of $\mathcal{B}$ is obvious since both Φ and i are linear.

(*ii*) Suppose $\{f_n\}$ is a bounded sequence in $L^2(\Omega)$. Then, $\Phi(f_n) = u_{f_n}$. By the Lax–Milgram theorem and (14), $\|u_{f_n}\|_{H_\gamma} \leq C \|f_n\|_{L^2(\Omega)}$ so that $\{\Phi(f_n)\}$ is bounded in H_γ. Since i is compact, $\{i \circ \Phi(f_n)\}$ has a convergent subsequence in $L^2(\Omega)$. Hence, $\mathcal{B}$ is compact.

(*iii*) Using (13) and the symmetry of A, we have

$$\langle \mathcal{B}f, g \rangle_{L^2(\Omega) \times L^2(\Omega)} = \int_\Omega u_f g \, dx = \int_\Omega g u_f \, dx$$

$$= \int_{\Omega \setminus \Gamma} A \nabla u_g \nabla u_f \, dx + \int_\Gamma h(u_{1g} - u_{2g})(u_{1f} - u_{2f}) \, d\sigma$$

$$= \int_{\Omega \setminus \Gamma} A \nabla u_f \nabla u_g \, dx + \int_\Gamma h(u_{1f} - u_{2f})(u_{1g} - u_{2g}) \, d\sigma$$

$$= \int_\Omega f u_g \, dx = \langle f, \mathcal{B}g \rangle_{L^2(\Omega) \times L^2(\Omega)}.$$

Therefore, the map $\mathcal{B}$ is symmetric.

(*iv*) Finally, by uniqueness (see Remark 2), $u_f = 0$ is the only solution of problem (13) if $f = 0$. Thus, we deduce statement (*iv*) of (12).

It follows that $\mathcal{B}$ satisfies the assumptions of Theorem 2 so that, in particular, conclusion (*ii*) of Theorem 2 follows. Also, there exist a decreasing sequence of the eigenvalues $\{\rho_k\}$ (counting the multiplicity) that goes to zero and a sequence of corresponding eigenvectors $\{u_k\}$, which verifies that $\mathcal{B}u_k = \rho_k u_k$ and forms a complete orthonormal system in $L^2(\Omega)$. Finally, observe that

$$\int_{\Omega \setminus \Gamma} A \nabla u_k \nabla v \, dx + \int_\Gamma h(u_{1k} - u_{2k})(v_1 - v_2) \, d\sigma = \lambda_k \int_\Omega u_k v \, dx$$

is equivalent to

$$\int_{\Omega \setminus \Gamma} A \nabla \left(\frac{1}{\lambda_k} u_k \right) \nabla v \, dx + \int_\Gamma h \left(\left(\frac{1}{\lambda_k} u_{1k} \right) - \left(\frac{1}{\lambda_k} u_{2k} \right) \right) (v_1 - v_2) \, d\sigma = \int_\Omega u_k v \, dx.$$

Thus, by the definition of the operator in (15), $\mathcal{B}u_k = \dfrac{1}{\lambda_k}u_k$. Therefore, for each $k \in \mathbb{N}$, $\lambda_k = \dfrac{1}{\rho_k}$ implies that $\{\lambda_k\}$ is an increasing sequence that goes to $+\infty$. The fact that the eigenvalues of problem (9) are positive follows from the coerciveness of the matrix A in (2) and assumption (5) of the function h. Indeed,

$$\lambda \int_\Omega u^2 \, dx = \int_{\Omega \backslash \Gamma} A \nabla u \nabla u \, dx + \int_\Gamma h(u_1 - u_2)^2 \, d\sigma$$

$$\geq \alpha \|\nabla u\|^2_{L^2(\Omega \backslash \Gamma)} + h_0 \|u_1 - u_2\|^2_{L^2(\Gamma)} > 0. \tag{16}$$

This completes the proof. $\qquad\square$

In the following proposition, we derive some characterizations of the eigenvalues of problem (9). They are obtained by adapting to our problem the classical principles on the characterization of eigenvalues (see [8]).

Proposition 2 *For $\gamma \leq 1$ and fixed ε, let $\{\lambda_\ell^\varepsilon\}$ be the sequence of the eigenvalues of problem (9) and $\{u_\ell^\varepsilon\}$ the sequence of the corresponding eigenvectors.*
Then, for $\ell \geq 1$,

$$\lambda_\ell^\varepsilon = \max_{\substack{v \in V_\ell \\ \|v\|_{L^2(\Omega)}=1}} \left(\int_{\Omega \backslash \Gamma^\varepsilon} A^\varepsilon \nabla v \nabla v \, dx + \varepsilon^\gamma \int_{\Gamma^\varepsilon} h^\varepsilon (v_1 - v_2)^2 \, d\sigma \right) \tag{17}$$

$$= \min_{\substack{v \in V_{\ell-1}^\perp \\ \|v\|_{L^2(\Omega)}=1}} \left(\int_{\Omega \backslash \Gamma^\varepsilon} A^\varepsilon \nabla v \nabla v \, dx + \varepsilon^\gamma \int_{\Gamma^\varepsilon} h^\varepsilon (v_1 - v_2)^2 \, d\sigma \right) \tag{18}$$

$$= \min_{M \in D_\ell^\varepsilon} \max_{\substack{v \in M \\ \|v\|_{L^2(\Omega)}=1}} \left(\int_{\Omega \backslash \Gamma^\varepsilon} A^\varepsilon \nabla v \nabla v \, dx + \varepsilon^\gamma \int_{\Gamma^\varepsilon} h^\varepsilon (v_1 - v_2)^2 \, d\sigma \right) \tag{19}$$

$$= \max_{M \in D_{\ell-1}^\varepsilon} \min_{\substack{v \in M^\perp \\ \|v\|_{L^2(\Omega)}=1}} \left(\int_{\Omega \backslash \Gamma^\varepsilon} A^\varepsilon \nabla v \nabla v \, dx + \varepsilon^\gamma \int_{\Gamma^\varepsilon} h^\varepsilon (v_1 - v_2)^2 \, d\sigma \right), \tag{20}$$

where $V_\ell = span\{u_1^\varepsilon, \ldots, u_\ell^\varepsilon\}$ and $D_\ell^\varepsilon = \{M \subset H_\gamma^\varepsilon | \dim M = \ell\}$.

Proof As in the previous proof, we omit ε in the notations. In view of Theorem 1, the sequence $\{u_\ell\}$ forms a complete orthonormal system in $L^2(\Omega)$. Hence, any $v \in H_\gamma$ can be written as $v = \sum_{k=1}^{+\infty} c_k u_k$, with $c_k = \langle u_k, v \rangle_{L^2(\Omega) \times L^2(\Omega)}$.

It is easy to verify that due to the orthonormality of the sequence $\{u_\ell\}$, if $\|v\|_{L^2(\Omega)} = 1$, then $\sum_{k=1}^{+\infty} |c_k|^2 = 1$. Now, define the map $a : (u, v) \in H_\gamma \times H_\gamma \longmapsto \mathbb{R}$ as

$$a(u, v) = \int_{\Omega \backslash \Gamma} A \nabla u \nabla v \, dx + \int_\Gamma h(u_1 - u_2)(v_1 - v_2) \, d\sigma.$$

Observe that a is a bilinear form on H_γ from the properties of integration and differentiation. Moreover,

$$a(v, v) = a \left(\sum_{k=1}^{+\infty} c_k u_k, \sum_{k=1}^{+\infty} c_k u_k \right) = \sum_{i,j=1}^{+\infty} c_i c_j \, a(u_i, u_j)$$

$$= \sum_{i,j=1}^{+\infty} c_i c_j \, \lambda_i \left\langle u_i, u_j \right\rangle_{L^2(\Omega) \times L^2(\Omega)}$$

$$= \sum_{i,j=1}^{+\infty} c_i c_j \, \lambda_i \delta_{ij} = \sum_{k=1}^{+\infty} |c_k|^2 \, \lambda_k. \tag{21}$$

Let $\mathcal{V}_\ell = \text{span}\{u_1, u_2, \ldots, u_\ell\}$.

We first show (17). For all $v \in \mathcal{V}_\ell$ such that $\|v\|_{L^2(\Omega)} = \sum_{k=1}^\ell |c_k|^2 = 1$, using (10) and (21), we deduce that

$$a(v, v) = \sum_{k=1}^\ell |c_k|^2 \, \lambda_k \leq \lambda_\ell \sum_{k=1}^\ell |c_k|^2 = \lambda_\ell.$$

Hence, $\lambda_\ell = \max\limits_{\substack{v \in \mathcal{V}_\ell \\ \|v\|_{L^2(\Omega)}=1}} a(v, v)$. The maximum is attained at $v = \dfrac{u_\ell}{\|u_\ell\|_{L^2(\Omega)}}$. Indeed,

$$a \left(\frac{u_\ell}{\|u_\ell\|_{L^2(\Omega)}}, \frac{u_\ell}{\|u_\ell\|_{L^2(\Omega)}} \right) = \frac{1}{\|u_\ell\|_{L^2(\Omega)}^2} a(u_\ell, u_\ell)$$

$$= \frac{1}{\|u_\ell\|_{L^2(\Omega)}^2} \lambda_\ell \|u_\ell\|_{L^2(\Omega)}^2 = \lambda_\ell. \tag{22}$$

Now, to prove (18), let $v \in \mathcal{V}_{\ell-1}^\perp$ such that $\|v\|_{L^2(\Omega)} = \sum_{k=\ell}^{+\infty} |c_k|^2 = 1$. From (10) and (21), we have

$$a(v, v) = \sum_{k=\ell}^{+\infty} |c_k|^2 \, \lambda_k \geq \lambda_\ell \sum_{k=\ell}^{+\infty} |c_k|^2 = \lambda_\ell.$$

Hence, $\lambda_\ell = \min\limits_{\substack{v \in \mathcal{V}_{\ell-1}^\perp \\ \|v\|_{L^2(\Omega)}=1}} a(v, v)$ where the minimum is attained at $v = \dfrac{u_\ell}{\|u_\ell\|_{L^2(\Omega)}}$

using the same argument as in (22).

Concerning the characterization in (19), let M be an ℓ-dimensional subspace of H_γ, and consider a basis $\{e_1, \ldots, e_\ell\}$ of this subspace. Hence, for any $v \in H_\gamma \backslash \{0\}$, we can write $v = \sum_{i=1}^{\ell} d_i e_i$, for some constants d_i, with at least one to be nonzero.

Let $\mathcal{V}_{\ell-1} = \text{span}\{u_1, u_2, \ldots, u_{\ell-1}\}$, and denote by $\mathcal{V}_{\ell-1}^\perp$ its orthogonal complement in H_γ. We choose $v_1 \in \mathcal{V}_{\ell-1}^\perp \backslash \{0\}$ such that $\|v_1\|_{L^2(\Omega)} = \sum_{k=\ell}^{+\infty} |c_k|^2 = 1$. This function exists since the system of equations $\sum_{i=1}^{\ell} d_i \langle e_i, u_j \rangle_{L^2(\Omega) \times L^2(\Omega)} = 0$ for $j = 1, .., \ell - 1$ has $\ell - 1$ equations with ℓ unknowns. Thus, a nontrivial vector of coefficients exists. As a result, $a(v_1, v_1) = \sum_{k=\ell}^{+\infty} |c_k|^2 \lambda_k \geq \lambda_\ell$. Hence, $\min\limits_{M \in \mathcal{D}_\ell} \max\limits_{\substack{v \in M \\ \|v\|_{L^2(\Omega)}=1}} a(v, v) \geq \lambda_\ell$. Finally, using (17), we get $\min\limits_{M \in \mathcal{D}_\ell} \max\limits_{\substack{v \in M \\ \|v\|_{L^2(\Omega)}=1}} a(v, v) \leq \lambda_\ell$, which proves the desired equality.

Now, we show the last characterization in (20). Let S be any $(\ell - 1)$-dimensional subspace of H_γ. There exists a function $w_1 \in S^\perp$ of the form $w_1 = \sum_{i=1}^{\ell} k_i u_i$. Indeed, this function exists since for any basis $\{\xi_1, \ldots, \xi_{\ell-1}\}$ of S, the system of equations

$$\sum_{i=1}^{\ell} k_i \langle u_i, \xi_j \rangle = 0 \text{ for } j = 1, .., \ell - 1 \tag{23}$$

has $\ell - 1$ equations with ℓ unknowns. Thus, a nontrivial vector of coefficients exists. Moreover, we can choose the coefficients k_i in (23) so that $\|w_1\|_{L^2(\Omega)} = \sum_{k=1}^{\ell} |k_i|^2 = 1$. As a result, $a(w_1, w_1) = \sum_{i=1}^{\ell} |k_i|^2 \lambda_k \leq \lambda_\ell$. Hence, $\max\limits_{M \in \mathcal{D}_{\ell-1}} \min\limits_{\substack{v \in M^\perp \\ \|v\|_{L^2(\Omega)}=1}} a(v, v) \leq \lambda_\ell$. From (19), we have $\max\limits_{M \in \mathcal{D}_{\ell-1}} \min\limits_{\substack{v \in M^\perp \\ \|v\|_{L^2(\Omega)}=1}} a(v, v) \geq \lambda_\ell$, which completes the proof. $\square$

3   Review of the Homogenization of an Elliptic Problem in a Two-Component Domain

In this section, we recall the main results of the periodic unfolding method applied to the homogenization of some elliptic problems in a two-component domain, investigated in [12] and [10].

3.1   The Periodic Unfolding Method in a Two-Component Domain

We briefly recall here the definitions and some properties of the unfolding operators $\mathcal{T}_1^\varepsilon$ and $\mathcal{T}_2^\varepsilon$ defined on Ω_1^ε and Ω_2^ε, respectively. We refer to [5] and [12] for the complete proofs and discussions.

Definition 2 For any Lebesgue-measurable function φ on Ω_i^ε, the unfolding operator $\mathcal{T}_i^\varepsilon$, for $i = 1, 2$, is defined as

$$
\mathcal{T}_i^\varepsilon(\varphi)(x, y) = \begin{cases} \varphi\left(\varepsilon\left[\dfrac{x}{\varepsilon}\right]_Y + \varepsilon y\right) & \text{a.e. for } (x, y) \in \widehat{\Omega}_\varepsilon \times Y_i, \\ 0 & \text{a.e. for } (x, y) \in \Lambda_\varepsilon \times Y_i, \end{cases}
$$

where $\widehat{\Omega}_\varepsilon$ is the union of the translated cells εY well contained in Ω and $\Lambda_\varepsilon = \Omega \setminus \widehat{\Omega}_\varepsilon$.

Note that in the definition above, if φ is a function defined in Ω, we simply write $\mathcal{T}_i^\varepsilon(\varphi)$ instead of $\mathcal{T}_i^\varepsilon(\varphi_{|\Omega_i^\varepsilon})$, $i = 1, 2$. Moreover, we define $\mathcal{T}_\varepsilon$ in $\Omega \times Y$ by the formula

$$
\mathcal{T}_\varepsilon(\varphi) = \begin{cases} \mathcal{T}_1^\varepsilon(\varphi) & \text{in } \Omega \times Y_1, \\ \mathcal{T}_2^\varepsilon(\varphi) & \text{in } \Omega \times Y_2. \end{cases}
$$

Proposition 3 ([12]) *For $p \in [1, +\infty[$, the operators $\mathcal{T}_i^\varepsilon$, $i = 1, 2$, are linear and continuous from $L^p(\Omega_i^\varepsilon)$ to $L^p(\Omega \times Y)$. Moreover,*

(i) $\mathcal{T}_i^\varepsilon(\varphi \psi) = \mathcal{T}_i^\varepsilon(\varphi)\mathcal{T}_i^\varepsilon(\psi)$ *for every Lebesgue-measurable functions φ and ψ on Ω_i^ε;*

(ii) *for every $\varphi \in L^1(\Omega_i^\varepsilon)$,*

$$
\frac{1}{|Y|} \int_{\Omega \times Y_i} \mathcal{T}_i^\varepsilon(\varphi)(x, y) \, dx \, dy = \int_{\Omega_i^\varepsilon} \varphi(x) \, dx - \int_{\Lambda^\varepsilon} \varphi(x) \, dx;
$$

moreover, if φ_ε is a bounded sequence in $L^{1+\eta}(\Omega_i^\varepsilon)$ for some $\eta > 0$, then

$$
\lim_{\varepsilon \to 0} \frac{1}{|Y|} \int_{\Omega \times Y_i} \mathcal{T}_i^\varepsilon(\varphi_\varepsilon)(x, y) \, dx \, dy = \lim_{\varepsilon \to 0} \int_{\Omega_i^\varepsilon} \varphi_\varepsilon(x) \, dx;
$$

(iii) $\|\mathcal{T}_i^\varepsilon(\varphi)\|_{L^p(\Omega \times Y_i)} \le |Y|^{\frac{1}{p}} \|\varphi\|_{L^p(\Omega_i^\varepsilon)}$ *for every $\varphi \in L^p(\Omega_i^\varepsilon)$;*

(iv) $\mathcal{T}_i^\varepsilon(\varphi) \to \varphi$ *strongly in $L^p(\Omega \times Y_i)$ for every $\varphi \in L^p(\Omega)$;*

(v) *if $\{\varphi_\varepsilon\}$ is a sequence in $L^p(\Omega)$ such that $\varphi_\varepsilon \to \varphi$ strongly in $L^p(\Omega)$, then $\mathcal{T}_i^\varepsilon(\varphi_\varepsilon) \to \varphi$ strongly in $L^p(\Omega \times Y_i)$;*

(vi) *if $\varphi \in L^p(Y_i)$ is Y-periodic and $\varphi_\varepsilon(x) = \varphi\left(\dfrac{x}{\varepsilon}\right)$, then $\mathcal{T}_i^\varepsilon(\varphi_\varepsilon) \to \varphi$ strongly in $L^p(\Omega \times Y_i)$;*

(vii) *if $\varphi_\varepsilon \in L^p(\Omega_i^\varepsilon)$ satisfies $\|\varphi_\varepsilon\|_{L^p(\Omega_i^\varepsilon)} \le C$ and $\mathcal{T}_i^\varepsilon(\varphi_\varepsilon) \rightharpoonup \widehat{\varphi}$ weakly in $L^p(\Omega \times Y_i)$, then $\widetilde{\varphi}_\varepsilon \rightharpoonup \theta_i \mathcal{M}_{Y_i}(\widehat{\varphi})$ weakly in $L^p(\Omega)$;*

(viii) *if $\varphi \in W^{1,p}(\Omega_i^\varepsilon)$, then $\nabla_y(\mathcal{T}_i^\varepsilon(\varphi)) = \varepsilon \mathcal{T}_i^\varepsilon(\nabla\varphi)$ and $\mathcal{T}_i^\varepsilon(\varphi) \in L^2(\Omega, W^{1,p}(Y_i))$.*

Remark 4 Suppose $u^\varepsilon = (u_1^\varepsilon, u_2^\varepsilon)$ is a bounded sequence in H_γ^ε, then for some C independent of ε, we have (see [12])

$$\left\| \mathcal{T}_i^\varepsilon(\nabla u_i^\varepsilon) \right\|_{L^2(\Omega \times Y_i)} \le C, i = 1, 2,$$

$$\left\| \mathcal{T}_1^\varepsilon(u_1^\varepsilon) - \mathcal{T}_2^\varepsilon(u_2^\varepsilon) \right\|_{L^2(\Omega \times \Gamma)} \le C\varepsilon^{\frac{1-\gamma}{2}}.$$

The following theorem presents the main convergence results for bounded sequences in H_γ^ε when $\gamma < 1$. We refer to [12] and [10] for the proof.

Theorem 3 ([12]) *Let* $\gamma < 1$ *and* $u^\varepsilon = (u_1^\varepsilon, u_2^\varepsilon)$ *be a bounded sequence in* H_γ^ε. *Then, there exist a subsequence (still denoted by* ε), $u_1 \in H_0^1(\Omega)$, $\widehat{u}_1 \in L^2(\Omega, H_{per}^1(Y_1))$, *and* $\widehat{u}_2 \in L^2(\Omega, H^1(Y_2))$ *such that*

$$
\begin{aligned}
\mathcal{T}_i^\varepsilon(u_i^\varepsilon) &\to u_1 && \text{strongly in } L^2(\Omega, H^1(Y_i)), i = 1, 2, \\
\widetilde{u}_i^\varepsilon &\rightharpoonup \theta_i u_1 && \text{weakly in } L^2(\Omega), i = 1, 2, \\
\mathcal{T}_1^\varepsilon(\nabla u_1^\varepsilon) &\rightharpoonup \nabla u_1 + \nabla_y \widehat{u}_1 && \text{weakly in } L^2(\Omega \times Y_1), \\
\mathcal{T}_2^\varepsilon(\nabla u_2^\varepsilon) &\rightharpoonup \nabla_y \widehat{u}_2 && \text{weakly in } L^2(\Omega \times Y_2).
\end{aligned}
\tag{24}
$$

3.2   Homogenization of a General Elliptic Problem

We recall here some known results on the homogenization, via the periodic unfolding method, of the problem

$$
\begin{cases}
-\text{div}\,(A^\varepsilon \nabla u_1^\varepsilon) = f_1^\varepsilon & \text{in } \Omega_1^\varepsilon, \\
-\text{div}\,(A^\varepsilon \nabla u_2^\varepsilon) = f_2^\varepsilon & \text{in } \Omega_2^\varepsilon, \\
A^\varepsilon \nabla u_1^\varepsilon \cdot n_1^\varepsilon = -A^\varepsilon \nabla u_2^\varepsilon \cdot n_2^\varepsilon & \text{on } \Gamma^\varepsilon, \\
-A^\varepsilon \nabla u_1^\varepsilon \cdot n_1^\varepsilon = \varepsilon^\gamma h^\varepsilon (u_1^\varepsilon - u_2^\varepsilon) & \text{on } \Gamma^\varepsilon, \\
u_1^\varepsilon = 0 & \text{on } \partial\Omega,
\end{cases}
\tag{25}
$$

where the matrix A and the function h satisfy assumptions (5) and (4).

We also suppose that

$$\widetilde{f_i^\varepsilon} \rightharpoonup f_i \quad \text{weakly in } L^2(\Omega) \qquad i = 1, 2, \tag{26}$$

for some $f_i \in L^2(\Omega)$, which in particular implies that

$$f^\varepsilon \rightharpoonup f_1 + f_2 \quad \text{weakly in } L^2(\Omega). \tag{27}$$

Problem (25) reduces to problem (3) when $f_i^\varepsilon = \lambda^\varepsilon u_i^\varepsilon, i = 1, 2$.
Its variational formulation is

$$\begin{cases} \text{Find } u^\varepsilon \in H_\gamma^\varepsilon \text{ such that} \\ \int_{\Omega \backslash \Gamma^\varepsilon} A^\varepsilon \nabla u^\varepsilon \nabla v \, dx + \varepsilon^\gamma \int_{\Gamma^\varepsilon} h^\varepsilon (u_1^\varepsilon - u_2^\varepsilon)(v_1 - v_2) \, d\sigma = \int_\Omega f^\varepsilon v \, dx \\ \text{for all } v \in H_\gamma^\varepsilon. \end{cases} \tag{28}$$

As proved by Hummel [14], the boundedness of the sequence $\{u_f^\varepsilon\}$ in H_γ^ε is not guaranteed for $\gamma > 1$. Here, we do not consider the case $\gamma = 1$, as discussed in Remark 5.

In the case where $f^\varepsilon \equiv f$, the homogenization for problem (28) was originally studied using Tartar's method in [11, 16], and [17], and then by the periodic unfolding method in [12] where an unfolded homogenized problem is also obtained (see also [10] for a more general nonlinear jump condition). The proofs of the homogenization results for problem (28) where the right-hand side depends on ε are straightforward extensions of those given in [12]. We only state the results below, following the notations therein (which are a bit different from those in [10]).

- **The case $\gamma \in (-1, 1)$**

Theorem 4 *Under assumptions (5), (4), and (26), let $\gamma \in (-1, 1)$ and $u^\varepsilon = (u_1^\varepsilon, u_2^\varepsilon)$ be the solution of problem (25). Then, there exist $u_1 \in H_0^1(\Omega)$ and $\widehat{u}_1 \in L^2(\Omega, H_{\text{per}}^1(Y_1))$ such that convergences (24) hold for the whole sequence with $\nabla_y \widehat{u}_2 = 0$.*

Moreover, the pair $(u_1, \widehat{u}_1)$ is the unique solution of the problem

$$\begin{cases} u_1 \in H_0^1(\Omega), \ \widehat{u}_1 \in L^2(\Omega, H_{\text{per}}^1(Y_1)) \text{ with } M_\Gamma(\widehat{u}_1) = 0 \text{ for a.e. } x \in \Omega, \\ \dfrac{1}{|Y|} \int_{\Omega \times Y_1} A(y)(\nabla u_1 + \nabla_y \widehat{u}_1)(\nabla \varphi + \nabla_y \Phi_1) \, dx dy = \int_\Omega (f_1 + f_2)(x)\varphi(x) \, dx \\ \text{for all } \varphi \in H_0^1(\Omega), \text{ for all } \Phi_1 \in L^2(\Omega, H_{\text{per}}^1(Y_1)). \end{cases} \tag{29}$$

- **The cases $\gamma < -1$ and $\gamma = -1$**

Theorem 5 *Under assumptions (5), (4), and (26), let $\gamma < -1$ and $u^\varepsilon = (u_1^\varepsilon, u_2^\varepsilon)$ be the solution of problem (25). Then, there exist $u_1 \in H_0^1(\Omega)$, $\widehat{u}_1 \in L^2(\Omega, H_{\text{per}}^1(Y_1))$, and $\widehat{u}_2 \in L^2(\Omega, H^1(Y_2))$ such that convergences (24) hold for the whole sequence.*

Moreover, the pair $(u_1, \widehat{u})$ *is the unique solution of the problem*

$$
\begin{cases}
u_1 \in H_0^1(\Omega), \ \widehat{u} \in L^2(\Omega, H_{\text{per}}^1(Y)) \ \text{with} \ \mathcal{M}_\Gamma(\widehat{u}) = 0 \ \text{for a.e.} \ x \in \Omega, \\
\dfrac{1}{|Y|} \displaystyle\int_{\Omega \times Y} A(y)(\nabla u_1 + \nabla_y \widehat{u})(\nabla \varphi + \nabla_y \Phi) \, dx dy = \int_\Omega (f_1 + f_2)(x)\varphi(x) \, dx \\
\text{for all} \ \varphi \in H_0^1(\Omega) \ \text{and for all} \ \Phi \in L^2(\Omega, H_{\text{per}}^1(Y)),
\end{cases}
$$

$$(30)$$

where $\widehat{u} \in L^2(\Omega, H_{\text{per}}^1(Y))$ *is the extension by periodicity (still denoted by* $\widehat{u}$*) of the function*

$$
\widehat{u}(\cdot, y) = \begin{cases}
\widehat{u}_1(\cdot, y), & y \in Y_1, \\
\widehat{u}_2(\cdot, y) - y_\Gamma \nabla u_1, & y \in Y_2,
\end{cases}
$$

and $y_\Gamma = y - \mathcal{M}_\Gamma(y)$.

Theorem 6 *Under assumptions* (5), (4), *and* (26), *let* $\gamma = -1$ *and* $u^\varepsilon = (u_1^\varepsilon, u_2^\varepsilon)$ *be the solution of problem* (25).

Then, there exist $u_1 \in H_0^1(\Omega)$, $\widehat{u}_1 \in L^2(\Omega, H_{\text{per}}^1(Y_1))$, *and* $\overline{u}_2 \in L^2(\Omega, H^1(Y_2))$ *such that convergences* (24) *hold for the whole sequence.*

Moreover, if we denote by $\overline{u}_2 \in L^2(\Omega, H^1(Y_2))$ *the function (extended by periodicity)*

$$
\overline{u}_2 = \widehat{u}_2 - y_\Gamma \nabla u_1 - \xi_\Gamma, \quad \text{for some} \ \xi_\Gamma \in L^2(\Omega),
$$

then $(u_1, \widehat{u}_1, \overline{u}_2)$ *is the unique solution of the problem*

$$
\begin{cases}
u_1 \in H_0^1(\Omega), \ \widehat{u}_1 \in L^2(\Omega, H_{\text{per}}^1(Y_1)) \ \text{with} \ \mathcal{M}_\Gamma(\widehat{u}_1) = 0, \ \overline{u}_2 \in L^2(\Omega, H^1(Y_2)), \\
\dfrac{1}{|Y|} \displaystyle\int_{\Omega \times Y_1} A(y)(\nabla u_1 + \nabla_y \widehat{u}_1)(\nabla \varphi + \nabla_y \Phi_1) \, dx dy \\
+ \dfrac{1}{|Y|} \displaystyle\int_{\Omega \times Y_2} A(y)(\nabla u_1 + \nabla_y \overline{u}_2)(\nabla \varphi + \nabla_y \Phi_2) \, dx dy \\
+ \dfrac{1}{|Y|} \displaystyle\int_{\Omega \times \Gamma} h(y)(\widehat{u}_1 - \overline{u}_2)(\Phi_1 - \Phi_2) \, dx d\sigma = \int_\Omega (f_1 + f_2)(x)\varphi(x) \, dx \\
\text{for all} \ \varphi \in H_0^1(\Omega) \ \text{and for all} \ \Phi_i \in L^2(\Omega, H_{\text{per}}^1(Y_i)), \quad i = 1, 2.
\end{cases}
$$

Now, we present the homogenized matrices corresponding to the three cases, $\gamma \in (-1, 1)$, $\gamma < -1$, and $\gamma = -1$, obtained in [11, 16] and [17] (see also [12]), as well as the homogenization problem in Ω.

We denote by A_γ^0 the constant homogenized matrix, defined as follows. This matrix is also symmetric, see Proposition 4 below.

1. If $\gamma \in (-1, 1)$, then

$$A_\gamma^0 \lambda = \theta_1 \mathcal{M}_{Y_1} (A \nabla w_\lambda),$$ (31)

with $w_\lambda \in H^1(Y_1)$ solution, for any $\lambda \in \mathbb{R}^N$, of

$$\begin{cases} -\text{div}\,(A \nabla w_\lambda) = 0 & \text{in } Y_1, \\ (A \nabla w_\lambda) \cdot n_1 = 0 & \text{on } \Gamma, \\ w_\lambda - \lambda \cdot y \text{ is } Y\text{-periodic}, \\ \mathcal{M}_{Y_1}(w_\lambda - \lambda \cdot y) = 0. \end{cases}$$

2. If $\gamma < -1$, then

$$A_\gamma^0 \lambda = \mathcal{M}_Y (A \nabla w_\lambda),$$ (32)

with $w_\lambda \in H^1(Y_1)$ solution, for any $\lambda \in \mathbb{R}^N$, of

$$\begin{cases} -\text{div}\,(A \nabla w_\lambda) = 0 & \text{in } Y, \\ w_\lambda - \lambda \cdot y \text{ is } Y\text{-periodic}, \\ \mathcal{M}_Y(w_\lambda - \lambda \cdot y) = 0. \end{cases}$$

3. If $\gamma = -1$, then

$$A_\gamma^0 = A_\gamma^1 + A_\gamma^2,$$ (33)

where

$$A_\gamma^i \lambda = \theta_i \mathcal{M}_{Y_i} \left(A \nabla w_\lambda^i \right), \quad i = 1, 2,$$

with $(w_\lambda^1, w_\lambda^2) \in H^1(Y_1) \times H^1(Y_2)$ solution, for any $\lambda \in \mathbb{R}^n$, of

$$\begin{cases} -\text{div}\,(A \nabla w_\lambda^i) = 0 \text{ in } Y_i, \ i = 1, 2 \\ A \nabla w_\lambda^1 \cdot n_1 = -A \nabla w_\lambda^2 \cdot n_2 \text{ on } \Gamma, \\ -A \nabla w_\lambda^1 \cdot n_1 = h(w_\lambda^1 - w_\lambda^2) \text{ on } \Gamma, \\ w_\lambda^1 - \lambda \cdot y \text{ is } Y\text{-periodic}, \\ \mathcal{M}_{Y_1}(w_\lambda^1 - \lambda \cdot y) = 0. \end{cases}$$

Theorem 7 *Let $\gamma < 1$. Under assumptions (5), (4), and (26), let $u^\varepsilon = (u_1^\varepsilon, u_2^\varepsilon)$ be the solution of problem (25). Then, there the $u_1 \in H_0^1(\Omega)$ such that*

$$\widetilde{u_1^\varepsilon} \rightharpoonup \theta_1 u_1 \qquad weakly\ in\ L^2(\Omega),$$

$$\widetilde{u_2^\varepsilon} \rightharpoonup \theta_2 u_1 \qquad weakly\ in\ L^2(\Omega),$$

$$A^\varepsilon \widetilde{\nabla u_1^\varepsilon} \rightharpoonup A_\gamma^0 \nabla u_1 \qquad weakly\ in\ \left(L^2(\Omega)\right)^N\ if\ \gamma \in (-1, 1),$$

$$A^\varepsilon \widetilde{\nabla u_2^\varepsilon} \rightharpoonup 0 \qquad weakly\ in\ \left(L^2(\Omega)\right)^N\ if\ \gamma \in (-1, 1),$$

$$A^\varepsilon \widetilde{\nabla u_1^\varepsilon} + A^\varepsilon \widetilde{\nabla u_2^\varepsilon} \rightharpoonup A_\gamma^0 \nabla u_1 \qquad weakly\ in\ \left(L^2(\Omega)\right)^N\ if\ \gamma \leq -1.$$

The function u_1 is the unique solution of the problem

$$\begin{cases} -\mathrm{div}\,(A_\gamma^0 \nabla u_1) = f_1 + f_2, & in\ \Omega \\ \qquad\qquad u_1 = 0 & on\ \partial\Omega, \end{cases} \qquad (34)$$

where A_γ^0 is defined in (31), (32), and (33) for the cases, $\gamma \in (-1, 1)$, $\gamma < -1$, and $\gamma = -1$, respectively.

The next proposition, adapted from [2] and [3], is important in the spectrum analysis of the homogenized problem.

Proposition 4 *Suppose A is symmetric. Then, the matrix A_γ^0 defined in (31), (32), and (33) for the cases, $\gamma \in (-1, 1)$, $\gamma < -1$, and $\gamma = -1$, respectively, is symmetric too. Moreover, $A_\gamma^0 \in M(\alpha_0, \beta_0, \Omega)$ for some positive number α_0 and $\beta_0 = \max_{i,j} |a_{ij}^0|$.*

4 Homogenization of an Elliptic Eigenvalue Problem in a Two-Component Domain

In this section, we describe the asymptotic behaviour of the eigenvalues and eigenvectors satisfying problem (3). Using the homogenization results given in the previous section, to prove the results we adapt to our problem some ideas introduced in [15] for the case of a fixed domain and [18] for that of perforated domains (see also [2]). An essential step to this aim is to prove some a priori estimates of the eigenvalues and eigenvectors.

We first consider the eigenvalue problem associated with the limit problem (34):

$$
\begin{cases}
-\mathrm{div}(A_\gamma^0 \nabla u) = \lambda u & \text{in } \Omega \\
u = 0 & \text{on } \partial\Omega,
\end{cases}
\tag{35}
$$

where A_γ^0 is a symmetric matrix and belongs to $M(\alpha_0, \beta_0, \Omega)$, for some $\alpha_0, \beta_0 \in \mathbb{R}$ (see Proposition 4).

The variational formulation of problem (35) is given by

$$
\begin{cases}
\text{Find } u \in H_0^1(\Omega) \text{ such that for each } v \in H_0^1(\Omega), \\
\displaystyle\int_\Omega A_\gamma^0 \nabla u \nabla v \, dx = \lambda \int_\Omega u v \, dx.
\end{cases}
\tag{36}
$$

This is similar to problem (9), but now posed in the whole domain Ω.

So, if we let $A = A_\gamma^0$ in Theorem 1, we deduce analogous results concerning the sequence of the eigenvalues in increasing order, denoted $\{\lambda_\ell\}$, where each eigenvalue is repeated as many times as the dimension of its corresponding eigenspace, as well as the sequence of corresponding eigenvectors $\{u_\ell\}$ in $H_0^1(\Omega)$, which forms a complete orthonormal system in $L^2(\Omega)$.

Furthermore, we also derive similar characterizations of the eigenvalues in Proposition 2.

That is,

$$
\lambda_\ell = \max_{\substack{v \in W_\ell \\ \|v\|_{L^2(\Omega)}=1}} \int_\Omega A_\gamma^0 \nabla v \nabla v \, dx = \min_{\substack{v \in W_{\ell-1}^\perp \\ \|v\|_{L^2(\Omega)}=1}} \int_\Omega A_\gamma^0 \nabla v \nabla v \, dx
\tag{37}
$$

$$
= \min_{W \in D_\ell} \max_{\substack{v \in W \\ \|v\|_{L^2(\Omega)}=1}} \int_\Omega A_\gamma^0 \nabla v \nabla v \, dx = \max_{W \in D_{\ell-1}} \min_{\substack{v \in W^\perp \\ \|v\|_{L^2(\Omega)}=1}} \int_\Omega A_\gamma^0 \nabla v \nabla v \, dx,
\tag{38}
$$

where $W_\ell = \mathrm{span}\{u_1, \ldots, u_\ell\}$ and $D_\ell = \{W \subset H_0^1 \mid \dim W = \ell\}$.

4.1 A Priori Estimates

In Theorem 1, we characterized the eigenvalues and eigenvectors corresponding to problem (3). We will now obtain some a priori estimates of these eigenvalues and eigenvectors. Under assumptions (5) and (26), denote by $\{\lambda_\ell^\varepsilon\}$ the sequence of the eigenvalues and $\{u_\ell^\varepsilon\}$ the sequence of the corresponding eigenvectors of the problem

$$
\begin{cases}
-\operatorname{div}\,(A^\varepsilon \nabla u_{1\ell}^\varepsilon) = \lambda_\ell^\varepsilon u_{1\ell}^\varepsilon & \text{in } \Omega_1^\varepsilon, \\[4pt]
-\operatorname{div}\,(A^\varepsilon \nabla u_{2\ell}^\varepsilon) = \lambda_\ell^\varepsilon u_{2\ell}^\varepsilon & \text{in } \Omega_2^\varepsilon, \\[4pt]
A^\varepsilon \nabla u_{1\ell}^\varepsilon \cdot n_1^\varepsilon = -A^\varepsilon \nabla u_{2\ell}^\varepsilon \cdot n_2^\varepsilon & \text{on } \Gamma^\varepsilon, \\[4pt]
-A^\varepsilon \nabla u_{1\ell}^\varepsilon \cdot n_1^\varepsilon = \varepsilon^\gamma h^\varepsilon (u_{1\ell}^\varepsilon - u_{2\ell}^\varepsilon) & \text{on } \Gamma^\varepsilon, \\[4pt]
u_{1\ell}^\varepsilon = 0 & \text{on } \partial\Omega, \\[4pt]
\|u_\ell^\varepsilon\|_{L^2(\Omega)} = 1.
\end{cases}
\tag{39}
$$

Observe that by construction (and for efficiency in the computations), the sequence $\{u_\ell^\varepsilon\}$ is a complete orthonormal system in $L^2(\Omega)$.

Proposition 5 *Let $\gamma < 1$ and $\{\lambda_\ell^\varepsilon\}$ be the sequence of the eigenvalues of Problem (39). Let $\{\lambda_\ell\}$ be the sequence of the eigenvalues of the homogenized problem (36) with coefficient matrix A_γ^0 given by (31)–(33) for the cases $\gamma \in (-1, 1)$, $\gamma < -1$, and $\gamma = -1$, respectively. Then, for a fixed ℓ, there exists a constant $C_{\lambda_\ell} > 0$ (depending on λ_ℓ), independent of ε, such that*

$$
\lambda_\ell^\varepsilon \le C_{\lambda_\ell}.
\tag{40}
$$

Proof Let $D_{\gamma,\ell}^\varepsilon = \{V \subset H_\gamma^\varepsilon \mid \dim V = \ell\}$ and $W_\ell = \operatorname{span}\{u_1, u_2, \ldots, u_\ell\}$, where $\{u_\ell\}$ is the sequence of the corresponding eigenvectors of the homogenized problem (36). Observe that for every ε, the subspace W_ℓ belongs to $D_{\gamma,\ell}^\varepsilon$ and its elements have no jump on Γ^ε.

Then, using Propositions 2, 4, (37), and (38), we have

$$
\lambda_\ell^\varepsilon = \min_{\substack{V \in D_{\gamma,\ell}^\varepsilon}} \max_{\substack{v \in V \\ \|v\|_{L^2(\Omega)}=1}} \left(\int_{\Omega \setminus \Gamma^\varepsilon} A^\varepsilon \nabla v \nabla v \, dx + \varepsilon^\gamma \int_{\Gamma^\varepsilon} h^\varepsilon (v_1 - v_2)^2 \, d\sigma \right)
$$

$$
\le \max_{\substack{v \in W_\ell \\ \|v\|_{L^2(\Omega)}=1}} \int_\Omega A^\varepsilon \nabla v \nabla v \, dx \le \max_{\substack{v \in W_\ell \\ \|v\|_{L^2(\Omega)}=1}} \beta \int_\Omega |\nabla v|^2 \, dx
$$

$$
\le \max_{\substack{v \in W_\ell \\ \|v\|_{L^2(\Omega)}=1}} \frac{\beta}{\alpha_0} \int_\Omega A_\gamma^0 \nabla v \nabla v \, dx = \frac{\beta}{\alpha_0} \lambda_\ell = C_{\lambda_\ell}.
$$

$\square$

Concerning the eigenvectors $\{u_\ell^\varepsilon\}$, we have the following proposition:

Proposition 6 *Let $\gamma < 1$ and $\{u_\ell^\varepsilon\}$ be the sequence of the eigenvectors corresponding to the eigenvalues $\{\lambda_\ell^\varepsilon\}$ of problem (39). Then, for a fixed ℓ, there exists a constant $c_{\lambda_\ell} > 0$ (depending on λ_ℓ), independent of ε, such that $\left\|u_\ell^\varepsilon\right\|_{H_\gamma^\varepsilon}^2 \leq c_{\lambda_\ell}$.*

Proof From the coercivity of A (as in (16)), we observe that

$$\lambda_\ell^\varepsilon \left\|u_\ell^\varepsilon\right\|_{L^2(\Omega)}^2 \geq \alpha \left\|\nabla u_\ell^\varepsilon\right\|_{L^2(\Omega\backslash\Gamma^\varepsilon)}^2 + \varepsilon^\gamma h_0 \left\|u_{1\ell}^\varepsilon - u_{2\ell}^\varepsilon\right\|_{L^2(\Gamma^\varepsilon)}^2 .$$

Taking $C_1 = \min\{\alpha, h_0\}$, using (40) and the fact that $\left\|u_\ell^\varepsilon\right\|_{L^2(\Omega)}^2 = 1$, we get

$$\|u\|_{H_\gamma^\varepsilon}^2 \leq \frac{\lambda_\ell^\varepsilon}{C_1} \leq c_{\lambda_\ell},$$

where $c_{\lambda_\ell} = \dfrac{C_{\lambda_\ell}}{C_1}$. $\qquad\square$

As a consequence of Propositions 5 and 6, together with Theorem 3, we have the following convergence results.

Corollary 2 *Under the assumptions of Proposition 5, there exist a subsequence (still denoted by ε) and $\Lambda_\ell \in \mathbb{R}^+$ such that*

$$\lambda_\ell^\varepsilon \to \Lambda_\ell. \tag{41}$$

Moreover, there exist $U_{1\ell} \in H_0^1(\Omega)$ and $\widehat{U}_{1\ell} \in L^2(\Omega, H^1(Y_1))$ such that (for the above subsequence)

$$\widetilde{u_{i\ell}^\varepsilon} \rightharpoonup \theta_i U_{1\ell} \qquad\qquad \text{weakly in } L^2(\Omega), \quad i = 1, 2,$$

$$\mathcal{T}_i^\varepsilon(u_{i\ell}^\varepsilon) \to U_{1\ell} \qquad\qquad \text{strongly in } L^2(\Omega, H^1(Y_i)), \quad i = 1, 2, \tag{42}$$

$$\mathcal{T}_1^\varepsilon(\nabla u_{1\ell}^\varepsilon) \rightharpoonup \nabla U_{1\ell} + \nabla_y \widehat{U}_{1\ell} \quad \text{weakly in } L^2(\Omega \times Y_1).$$

4.2 Convergence of Eigenvalues and Eigenspaces

In this section, we examine the limit behaviour of problem (39) when $\varepsilon \to 0$.

To do that, we apply the homogenization results in Sect. 3.2 with

$$f_i^\varepsilon = \lambda_\ell^\varepsilon u_{i\ell}^\varepsilon \quad i = 1, 2. \tag{43}$$

Observe from Corollary 2 that (up a subsequence) for every ℓ, there exists $U_{1\ell} \in H_0^1(\Omega)$ such that

$$\widetilde{f_i^\varepsilon} = \lambda_\ell^\varepsilon \widetilde{u_{i\ell}^\varepsilon} \rightharpoonup f_i := \Lambda_\ell \theta_i U_{1\ell} \quad \text{weakly in } L^2(\Omega), \quad i = 1, 2. \tag{44}$$

Theorem 8 *For $\gamma < 1$, let $\{\lambda_\ell^\varepsilon\}$ be the sequence of the eigenvalues of problem (39) and $\{u_\ell^\varepsilon\}$ the sequence of the corresponding eigenvectors. Let also $\{\lambda_\ell\}$ be the sequence of the eigenvalues of the homogenized problem (36) with A_γ^0 given by (31), (32), and (33) for the cases $\gamma \in (-1, 1)$, $\gamma < -1$, and $\gamma = -1$, respectively.*

Then, for each fixed ℓ,

(i) $\lambda_\ell^\varepsilon \to \lambda_\ell$;

(ii) the eigenspaces of the ε-problem converge to the corresponding ones of the homogenized problem; namely, there exists a subsequence (still denoted by ε) such that

$$
\begin{aligned}
&(a) \; u_\ell^\varepsilon \to U_{1\ell} && \text{strongly in } \; L^2(\Omega),\\[4pt]
&(b) \; \mathcal{T}_i^\varepsilon(u_{i\ell}^\varepsilon) \to U_{1\ell} && \text{strongly in } \; L^2(\Omega, H^1(Y_i)), \quad i = 1, 2,\\[4pt]
&(c) \; \mathcal{T}_1^\varepsilon(\nabla u_{1\ell}^\varepsilon) \rightharpoonup \nabla U_{1\ell} + \nabla_y \widehat{U}_{1\ell} && \text{weakly in } \; L^2(\Omega \times Y_1),\\[4pt]
&(d) \; \mathcal{T}_2^\varepsilon(\nabla u_{2\ell}^\varepsilon) \rightharpoonup 0 && \text{weakly in } \; L^2(\Omega \times Y_2).
\end{aligned}
$$

$$(45)$$

Moreover, $U_{1\ell}$ is an eigenvector of the homogenized problem (36) corresponding to λ_ℓ, and $\{U_{1\ell}\}$ forms a complete orthonormal system in $L^2(\Omega)$;

(iii) if the homogenized eigenvalue λ_ℓ is simple, then that one can choose for every ε the eigenfunction u_ℓ^ε such as the whole sequence $\{u_\ell^\varepsilon\}$ converges to $U_{1\ell}$.

Remark 5 No convergence result for the eigenvalue problem has been obtained for the case $\gamma = 1$, although homogenization results exist in literature, because in that case the unfolding of the second component u_2^ε only converges weakly, so that u^ε does not converge strongly in $L^2(\Omega)$. Hence, the difficulty lies in showing that the sequence of homogenized eigenvectors is orthonormal in $L^2(\Omega)$, since we cannot argue as in the case of $\gamma < 1$ (see the proof of convergence (48) in the next section).

Let us mention that some results for the case $\gamma = 1$ were done by I. Gruais, D. Poliševski, and A. Ştefan in [14], where the convergence of the eigenvalues and eigenvectors is examined using the two-scale convergence method. However, the two-scale result (Theorem 3.1) stated there (without proof) is false as far as it concerns the disconnected component, as can be seen by counterexamples. Consequently, we have doubts on the veracity of their eigenvalue convergence results, which are deduced from Theorem 3.1.

5 Proof of Theorem 8

The weak convergence in $(ii\,a)$ and the convergences in $(ii\,b)$-$(ii\,c)$ in (45) are given by Corollary 2. On the other hand, $(ii\,d)$ can be proved similarly as in Theorem 4 with f^ε defined by (43).

Let us show (i) and the last sentence in (ii). This will also imply that convergence ($ii\ a$) is strong, since both norms, of u_ℓ^ε and of $U_{1\ell}$, are equal to 1, which implies the convergence of the norm.

From (44), we have

$$f_1 + f_2 = \Lambda_\ell(\theta_1 U_{1\ell} + \theta_2 U_{1\ell}) = \Lambda_\ell U_{1\ell}.$$

Hence, we deduce from Theorem 7 that $U_{1\ell}$ satisfies

$$\begin{cases} -\mathrm{div}\,(A_\gamma^0 \nabla U_{1\ell}) = \Lambda_\ell U_{1\ell} & \text{in } \Omega \\ U_{1\ell} = 0 & \text{on } \partial\Omega. \end{cases} \tag{46}$$

Now, we recall that from Theorem 1, the sequence $\{u_\ell^\varepsilon\}$ in H_γ^ε is orthonormal, that is,

$$\int_\Omega u_i^\varepsilon u_j^\varepsilon \, dx = \delta_{ij} \quad \forall i, j = 1, \ldots, N. \tag{47}$$

We use the unfolding method to obtain the limit of the integral as $\varepsilon \to 0$. We write

$$\int_\Omega u_i^\varepsilon u_j^\varepsilon \, dx = \int_{\Omega_1^\varepsilon} u_{1i}^\varepsilon u_{1j}^\varepsilon \, dx + \int_{\Omega_2^\varepsilon} u_{2i}^\varepsilon u_{2j}^\varepsilon \, dx.$$

Observe now that in view of Proposition 6, we can apply (ii) of Proposition 3, which using convergence ($ii\ b$) of (45) gives

$$\lim_{\varepsilon \to 0} \int_\Omega u_i^\varepsilon u_j^\varepsilon \, dx = \lim_{\varepsilon \to 0} \frac{1}{|Y|} \int_{\Omega \times Y_1} \mathcal{T}_1^\varepsilon(u_{1i}^\varepsilon)\mathcal{T}_1^\varepsilon(u_{1j}^\varepsilon) \, dxdy$$

$$+ \lim_{\varepsilon \to 0} \int_{\Omega \times Y_2} \mathcal{T}_2^\varepsilon(u_{2i}^\varepsilon)\mathcal{T}_2^\varepsilon(u_{2j}^\varepsilon) \, dxdy$$

$$= \frac{1}{|Y|} \int_{\Omega \times Y_1} U_{1i}U_{1j} \, dxdy + \frac{1}{|Y|} \int_{\Omega \times Y_2} U_{1i}U_{1j} \, dxdy \tag{48}$$

$$= \frac{|Y_1|}{|Y|} \int_\Omega U_{1i}U_{1j} \, dx + \frac{|Y_2|}{|Y|} \int_\Omega U_{1i}U_{1j} \, dx = \int_\Omega U_{1i}U_{1j} \, dx,$$

since U_{1i} is independent of y for every i. Hence,

$$\int_\Omega U_{1i}U_{1j} \, dx = \delta_{ij}, \qquad \text{for every } i, j, \tag{49}$$

so that the sequence $\{U_{1\ell}\}$ is orthonormal in $L^2(\Omega)$. This shows that $U_{1\ell} \neq 0$, for all $\ell \in \mathbb{N}$. Thus, $U_{1\ell}$ is an eigenvector corresponding to the eigenvalue Λ_ℓ of the homogenized problem (46). This also implies that these functions are linearly independent.

Recall from Corollary 2 that $\lambda_\ell^\varepsilon \to \Lambda_\ell$ up to a subsequence for some $\Lambda_\ell \in \mathbb{R}^+$. To end the proof, we now show that there are no other eigenvalues and corresponding eigenvectors except those defined by (41), (46), and (49), implying the uniqueness of the limit Λ_ℓ for all subsequences of $\{\lambda_\ell^\varepsilon\}$. It will follow that the convergence holds true for the whole sequence.

We give a proof by contradiction. Suppose that there is an eigenvector w corresponding to some eigenvalue Λ satisfying

$$
\begin{cases}
-\text{div}\,(A_\gamma^0 \nabla w) = \Lambda w & \text{in } \Omega \\
w = 0 & \text{on } \partial\Omega,
\end{cases}
\tag{50}
$$

and which is not given by (41), (46), and (49). Then, w does not belong to any subspace generated by a finite family of $\{U_{1\ell}\}$. Indeed, suppose otherwise that $w = \sum_{i=1}^m c_i U_{1i}$, where $c_i \neq 0$ are constants.

Then, using (46) and (50),

$$
\Lambda \sum_{i=1}^m c_i U_{1i} = \Lambda w = -\text{div}(A_\gamma^0 \nabla w) = -\text{div}\left(A_\gamma^0 \sum_{i=1}^m c_i \nabla U_{1i} \right)
$$

$$
= \sum_{i=1}^m c_i \left(-\text{div}\left(A_\gamma^0 \nabla U_{1i} \right) \right) = \sum_{i=1}^m c_i \Lambda_i U_{1i}.
$$

Thus, $\sum_{i=1}^m (\Lambda - \Lambda_i) c_i U_{1i} = 0$. This is only true when $\Lambda = \Lambda_i$, for all $i = 1, .., m$, because the functions U_{1i} are linearly independent. This is not possible due to the assumption on w. As a result, since w does not belong to any subspace generated by a finite family of $\{U_{1\ell}\}$, w is orthogonal to these sets.

Now, from (10) and (41), we deduce that $0 < \Lambda_1 \leq \Lambda_2 \leq \ldots \to +\infty$. Hence, there exists a ℓ_0 such that

$$
\Lambda_{\ell_0+1} > \Lambda.
\tag{51}
$$

The strict inequality is because of the assumption imposed on Λ.

In the following part, we will show that if such pair (Λ, w) exists, we will obtain the fact that $\Lambda_{\ell_0+1} \leq \Lambda$, which contradicts (51).

At this point, we introduce U^ε to be a solution of the following problem:

$$
\begin{cases}
-\text{div}\,(A^\varepsilon \nabla U^\varepsilon) = \Lambda w & \text{in } \Omega \backslash \Gamma^\varepsilon \\
A^\varepsilon \nabla U_1^\varepsilon n_1^\varepsilon = -A^\varepsilon \nabla U_2^\varepsilon n_2^\varepsilon & \text{on } \Gamma^\varepsilon \\
-A^\varepsilon \nabla U_1^\varepsilon n_1^\varepsilon = \varepsilon^\gamma h^\varepsilon (U_1^\varepsilon - U_2^\varepsilon) & \text{on } \Gamma^\varepsilon \\
U_1^\varepsilon = 0 & \text{on } \partial\Omega,
\end{cases}
\tag{52}
$$

where A and h satisfy assumptions (5) and (4).

Then, from Theorems 3 and 7 with $f^\varepsilon = \Lambda w$, there exists $U^0 \in H_0^1(\Omega)$ such that

$$\widetilde{U}_i^\varepsilon \rightharpoonup \theta_i U^0 \quad \text{weakly in } L^2(\Omega), \ i = 1, 2,$$
$$\mathcal{T}_i^\varepsilon(U_i^\varepsilon) \to U^0 \quad \text{strongly in } L^2(\Omega, H^1(Y_i)), \ i = 1, 2,$$
$$(53)$$

so that

$$U^\varepsilon \rightharpoonup U^0 \quad \text{weakly in } L^2(\Omega),$$

and U^0 is the unique solution of the homogenized problem

$$\begin{cases} -\text{div} \, (A_\gamma^0 \nabla U^0) = \Lambda w & \text{in } \Omega \\ \qquad\qquad U^0 = 0 & \text{on } \partial\Omega. \end{cases} \tag{54}$$

Problem (50) and the uniqueness of solution of (54) imply that

$$U^0 = w, \tag{55}$$

giving us

$$U^\varepsilon \rightharpoonup w \quad \text{weakly in } L^2(\Omega). \tag{56}$$

Recall from Theorem 1 that the sequence $\{u_\ell^\varepsilon\}$ forms a complete orthonormal system in $L^2(\Omega)$.

Hence, we can write

$$U^\varepsilon = \sum_{k=1}^{+\infty} c_k^\varepsilon u_k^\varepsilon, \tag{57}$$

where $c_k^\varepsilon = \langle U^\varepsilon, u_k^\varepsilon \rangle_{L^2(\Omega), L^2(\Omega)}$. Set

$$v^\varepsilon = U^\varepsilon - \sum_{k=1}^{\ell_0} c_k^\varepsilon u_k^\varepsilon = \sum_{k=\ell_0+1}^{+\infty} c_k^\varepsilon u_k^\varepsilon. \tag{58}$$

Following this construction, we have $\langle v^\varepsilon, u_k^\varepsilon \rangle_{L^2(\Omega), L^2(\Omega)} = 0$, for $k = 1, 2, \ldots, \ell_0$, which implies that $v^\varepsilon \in V_{\ell_0}^\perp$, where $V_{\ell_0} = \text{span}\{u_1^\varepsilon, \ldots, u_{\ell_0}^\varepsilon\}$.

From the characterization in Proposition 2, we have

$$\lambda_{\ell_0+1}^\varepsilon = \min_{\substack{v \in V_{\ell_0}^\perp \\ \|v\|_{L^2(\Omega)}=1}} \left(\int_{\Omega \setminus \Gamma^\varepsilon} A^\varepsilon \nabla v \nabla v \, dx + \varepsilon^\gamma \int_{\Gamma^\varepsilon} h^\varepsilon (v_1 - v_2)^2 \, d\sigma \right)$$

$$\leq \int_{\Omega \setminus \Gamma^\varepsilon} A^\varepsilon \nabla \left(\frac{v^\varepsilon}{\|v^\varepsilon\|_{L^2(\Omega)}} \right) \nabla \left(\frac{v^\varepsilon}{\|v^\varepsilon\|_{L^2(\Omega)}} \right) dx$$

$$+ \varepsilon^\gamma \int_{\Gamma^\varepsilon} h^\varepsilon \left(\frac{v_1^\varepsilon}{\|v^\varepsilon\|_{L^2(\Omega)}} - \frac{v_2^\varepsilon}{\|v^\varepsilon\|_{L^2(\Omega)}} \right)^2 d\sigma$$

$$= \frac{1}{\|v^\varepsilon\|_{L^2(\Omega)}^2} \left(\int_{\Omega \setminus \Gamma^\varepsilon} A^\varepsilon \nabla v^\varepsilon \nabla v^\varepsilon \, dx + \varepsilon^\gamma \int_{\Gamma^\varepsilon} h^\varepsilon (v_1^\varepsilon - v_2^\varepsilon)^2 \, d\sigma \right). \quad (59)$$

Passing to the limit in (59), we get

$$\Lambda_{\ell_0+1} \leq \lim_{\varepsilon \to 0} \frac{1}{\|v^\varepsilon\|_{L^2(\Omega)}^2} \left(\int_{\Omega \setminus \Gamma^\varepsilon} A^\varepsilon \nabla v^\varepsilon \nabla v^\varepsilon \, dx + \varepsilon^\gamma \int_{\Gamma^\varepsilon} h^\varepsilon (v_1^\varepsilon - v_2^\varepsilon)^2 \, d\sigma \right). \tag{60}$$

From the definition of v^ε in (58), we have

$$\int_{\Omega \setminus \Gamma^\varepsilon} A^\varepsilon \nabla v^\varepsilon \nabla v^\varepsilon \, dx = \int_{\Omega \setminus \Gamma^\varepsilon} A^\varepsilon \nabla U^\varepsilon \nabla U^\varepsilon \, dx - 2 \sum_{k=1}^{\ell_0} c_k^\varepsilon \int_{\Omega \setminus \Gamma^\varepsilon} A^\varepsilon \nabla U^\varepsilon \nabla u_k^\varepsilon \, dx$$

$$+ \sum_{k,j=1}^{\ell_0} c_k^\varepsilon c_j^\varepsilon \int_{\Omega \setminus \Gamma^\varepsilon} A^\varepsilon \nabla u_k^\varepsilon \nabla u_j^\varepsilon \, dx, \tag{61}$$

and

$$\varepsilon^\gamma \int_{\Gamma^\varepsilon} h^\varepsilon (v_1^\varepsilon - v_2^\varepsilon)^2 \, d\sigma$$

$$= \varepsilon^\gamma \int_{\Gamma^\varepsilon} h^\varepsilon \left[\left(U_1^\varepsilon - \sum_{k=1}^{\ell_0} c_k^\varepsilon u_{1k}^\varepsilon \right)^2 - 2 \left(U_1^\varepsilon - \sum_{k=1}^{\ell_0} c_k^\varepsilon u_{1k}^\varepsilon \right) \left(U_2^\varepsilon - \sum_{k=1}^{\ell_0} c_k^\varepsilon u_{2k}^\varepsilon \right) \right.$$

$$\left. + \left(U_2^\varepsilon - \sum_{k=1}^{\ell_0} c_k^\varepsilon u_{2k}^\varepsilon \right)^2 \right] d\sigma$$

$$= \quad \varepsilon^{\gamma} \int_{\Gamma^{\varepsilon}} h^{\varepsilon} (U_1^{\varepsilon} - U_2^{\varepsilon})^2 - 2 \sum_{k=1}^{\ell_0} c_k^{\varepsilon} \left[\varepsilon^{\gamma} \int_{\Gamma^{\varepsilon}} h^{\varepsilon} (U_1^{\varepsilon} - U_2^{\varepsilon})(u_{1k}^{\varepsilon} - u_{2k}^{\varepsilon}) \, d\sigma \right]$$

$$+ \quad \sum_{k,j=1}^{\ell_0} c_k^{\varepsilon} c_j^{\varepsilon} \left[\varepsilon^{\gamma} \int_{\Gamma^{\varepsilon}} h^{\varepsilon} (u_{1k}^{\varepsilon} - u_{2k}^{\varepsilon})(u_{1j}^{\varepsilon} - u_{2j}^{\varepsilon}) \, d\sigma \right]. \tag{62}$$

Using (61) and (62), we can write

$$\int_{\Omega \setminus \Gamma^{\varepsilon}} A^{\varepsilon} \nabla v^{\varepsilon} \nabla v^{\varepsilon} \, dx + \varepsilon^{\gamma} \int_{\Gamma^{\varepsilon}} h^{\varepsilon} (v_1^{\varepsilon} - v_2^{\varepsilon})^2 \, d\sigma$$

$$= \int_{\Omega \setminus \Gamma^{\varepsilon}} A^{\varepsilon} \nabla U^{\varepsilon} \nabla U^{\varepsilon} \, dx + \varepsilon^{\gamma} \int_{\Gamma^{\varepsilon}} h^{\varepsilon} (U_1^{\varepsilon} - U_2^{\varepsilon})^2 \, d\sigma$$

$$- 2 \sum_{k=1}^{\ell_0} c_k^{\varepsilon} \left[\int_{\Omega \setminus \Gamma^{\varepsilon}} A^{\varepsilon} \nabla U^{\varepsilon} \nabla u_k^{\varepsilon} \, dx + \varepsilon^{\gamma} \int_{\Gamma^{\varepsilon}} h^{\varepsilon} (U_1^{\varepsilon} - U_2^{\varepsilon})(u_{1k}^{\varepsilon} - u_{2k}^{\varepsilon}) \, d\sigma \right]$$

$$+ \sum_{k,j=1}^{\ell_0} c_k^{\varepsilon} c_j^{\varepsilon} \left[\int_{\Omega \setminus \Gamma^{\varepsilon}} A^{\varepsilon} \nabla u_k^{\varepsilon} \nabla u_j^{\varepsilon} \, dx + \varepsilon^{\gamma} \int_{\Gamma^{\varepsilon}} h^{\varepsilon} (u_{1k}^{\varepsilon} - u_{2k}^{\varepsilon})(u_{1j}^{\varepsilon} - u_{2j}^{\varepsilon}) \, d\sigma \right].$$

From the variational formulation of problems (39) and (52), we have

$$\int_{\Omega \setminus \Gamma^{\varepsilon}} A^{\varepsilon} \nabla v^{\varepsilon} \nabla v^{\varepsilon} \, dx + \varepsilon^{\gamma} \int_{\Gamma^{\varepsilon}} h^{\varepsilon} (v_1^{\varepsilon} - v_2^{\varepsilon})^2 \, d\sigma$$

$$= \Lambda \int_{\Omega} w U^{\varepsilon} \, dx - 2 \sum_{k=1}^{\ell_0} c_k^{\varepsilon} \Lambda \int_{\Omega} w u_k^{\varepsilon} \, dx + \sum_{k,j=1}^{\ell_0} c_k^{\varepsilon} c_j^{\varepsilon} \lambda_k^{\varepsilon} \int_{\Omega} u_k^{\varepsilon} u_j^{\varepsilon} \, dx$$

$$= \Lambda \int_{\Omega} w U^{\varepsilon} \, dx - 2 \sum_{k=1}^{\ell_0} c_k^{\varepsilon} \Lambda \int_{\Omega} w u_k^{\varepsilon} \, dx + \sum_{k=1}^{\ell_0} (c_k^{\varepsilon})^2 \lambda_k^{\varepsilon}. \tag{63}$$

On the other hand, using $(ii\,b)$ of (45) and (53), we deduce by unfolding that

$$c_k^{\varepsilon} = \int_{\Omega} U^{\varepsilon} u_k^{\varepsilon} \, dx = \int_{\Omega_1^{\varepsilon}} U_1^{\varepsilon} u_{1k}^{\varepsilon} \, dx + \int_{\Omega_2^{\varepsilon}} U_2^{\varepsilon} u_{2k}^{\varepsilon} \, dx$$

$$= \frac{1}{|Y|} \int_{\Omega \times Y_1} \mathcal{T}_1^{\varepsilon}(U_1^{\varepsilon}) \mathcal{T}_1^{\varepsilon}(u_{1k}^{\varepsilon}) \, dx dy + \int_{\Omega \times Y_2} \mathcal{T}_2^{\varepsilon}(U_2^{\varepsilon}) \mathcal{T}_2^{\varepsilon}(u_{2k}^{\varepsilon}) \, dx dy$$

$$\rightarrow \frac{1}{|Y|} \int_{\Omega \times Y_1} U^0 U_{1k} \, dx dy + \frac{1}{|Y|} \int_{\Omega \times Y_2} U^0 U_{1k} \, dx dy$$

$$= \frac{1}{|Y|} \int_\Omega U^0 U_{1k} \, dx \int_{Y_1} dy + \frac{1}{|Y|} \int_\Omega U^0 U_{1k} \, dx \int_{Y_2} dy$$

$$= \frac{|Y_1|}{|Y|} \int_\Omega U^0 U_{1k} \, dx + \frac{|Y_2|}{|Y|} \int_\Omega U^0 U_{1k} \, dx = \int_\Omega U^0 U_{1k} \, dx.$$

As a result, from (55), we have (up to a subsequence)

$$c_k^\varepsilon = \int_\Omega U^\varepsilon u_k^\varepsilon \, dx \to c_k := \int_\Omega w U_{1k} \, dx \quad \text{for all } k. \tag{64}$$

Now, let $W_{\ell_0} = \text{span}\{U_{11}, \ldots, U_{1\ell_0}\}$. Recall from the assumption of w that in particular, $w \perp W_{\ell_0}$. So,

$$c_k = 0, \quad \text{for } k = 1, .., \ell_0. \tag{65}$$

We then pass to the limit (up to a subsequence) at the right-hand side of (63). Using $(ii\,a)$ of (45), (41), (56), (64), and (65), we get

$$\int_{\Omega \setminus \Gamma^\varepsilon} A^\varepsilon \nabla v^\varepsilon \nabla v^\varepsilon \, dx + \varepsilon^\gamma \int_{\Gamma^\varepsilon} h^\varepsilon (v_1^\varepsilon - v_2^\varepsilon)^2 \, d\sigma$$

$$\to \Lambda \int_\Omega w^2 \, dx - 2 \sum_{k=1}^{\ell_0} c_k \Lambda \int_\Omega w U_{1k} \, dx + \sum_{k=1}^{\ell_0} c_k^2 \Lambda_k = \Lambda \int_\Omega w^2 \, dx. \tag{66}$$

On the other hand, using (58) and (47), we have

$$\|v^\varepsilon\|_{L^2(\Omega)}^2 = \int_\Omega (v^\varepsilon)^2 \, dx$$

$$= \int_\Omega (U^\varepsilon)^2 \, dx - 2 \sum_{k=1}^{\ell_0} c_k^\varepsilon \int_\Omega U^\varepsilon u_k^\varepsilon \, dx + \sum_{k,j=1}^{\ell_0} c_k^\varepsilon c_j^\varepsilon \int_\Omega u_k^\varepsilon u_j^\varepsilon \, dx$$

$$= \int_\Omega (U^\varepsilon)^2 \, dx - 2 \sum_{k=1}^{\ell_0} c_k^\varepsilon c_k^\varepsilon + \sum_{k,j=1}^{\ell_0} c_k^\varepsilon c_j^\varepsilon \delta_{kj}$$

$$= \int_\Omega (U^\varepsilon)^2 \, dx - \sum_{k=1}^{\ell_0} (c_k^\varepsilon)^2. \tag{67}$$

From (53), we see by unfolding that

$$\int_\Omega (U^\varepsilon)^2 \, dx \to \int_\Omega w^2 \, dx. \tag{68}$$

Using this result and (64)–(65), we then pass to the limit (up to a subsequence) in (67), and we deduce

$$\|v^\varepsilon\|^2_{L^2(\Omega)} \rightarrow \int_\Omega w^2 \, dx - \sum_{k=1}^{\ell_0} c_k^2 = \int_\Omega w^2 \, dx. \tag{69}$$

Using convergences (66) and (69) into (60), we obtain $\Lambda_{\ell_0+1} \leq \Lambda$, which contradicts (51). This proves that $\Lambda_\ell = \lambda_\ell$ for all ℓ, so that convergence (1) holds for the whole sequence and the sequence $\{\lambda_\ell\}$ contains all the eigenvalues of the problem and only them. This proves (1).

Moreover, the symmetry of the homogenized matrix A_γ^0 argued in Proposition 4 implies that the sequence $\{U_{1\ell}\}$ forms a complete orthonormal system in $L^2(\Omega)$, in view of Theorem 2.

Finally, we show item (iii). Let λ_ℓ be a simple eigenvalue and $U_{1\ell}$ be a corresponding eigenvector.

From (49), we have

$$\|U_{1\ell}\|^2_{L^2(\Omega)} = \int_\Omega U_{1\ell}^2 \, dx = 1. \tag{70}$$

Observe first that if the eigenvalue λ_ℓ is simple, then λ_ℓ^ε is simple too (for a sufficiently small ε). Then, for any $\varepsilon > 0$, let us denote by u_ℓ^ε the eigenvector corresponding to λ_ℓ^ε such that

$$\int_\Omega u_\ell^\varepsilon U_{1\ell} \, dx \geq 0. \tag{71}$$

Suppose that for a subsequence (still denoted by ε), we have $u_\ell^\varepsilon \rightharpoonup u_{1\ell}$ weakly in $L^2(\Omega)$, where $u_{1\ell}$ is another eigenvector corresponding to λ_ℓ.

Again from (49),

$$\|u_{1\ell}\|^2_{L^2(\Omega)} = \int_\Omega u_{1\ell}^2 \, dx = 1. \tag{72}$$

Consequently, we have two eigenvectors $U_{1\ell}$ and $u_{1\ell}$, associated with a simple eigenvalue λ_ℓ. Hence, there exists a constant C such that $U_{1\ell} = Cu_{1\ell}$. Together with (70) and (72), we get $|C| = 1$. Now, passing to the limit in (71), we obtain $\int_\Omega u_{1\ell} U_{1\ell} \, dx \geq 0$, so that $C = 1$. As a consequence,

$$U_{1\ell} = u_{1\ell}.$$

Thus, the whole sequence $\{u_\ell^\varepsilon\}$ converges to $U_{1\ell}$. This concludes the proof.

Acknowledgments This work was initiated during the appointment of the first author as a three-month Visiting Professor at the University of the Philippines (UP) Diliman in 2018. The support of this organization is greatly appreciated. Support from the University of Rouen Normandie for the first author's sabbatical leave is also acknowledged. The second author is grateful for the financial support of the UP System to attend the discussion meeting at the International Centre for Theoretical Sciences (ICTS), Bangalore. The authors are thankful to ICTS for the opportunity to meet and finalize the chapter.

References

1. Boccardo, L., Marcellini P.: Sulla convergenza delle soluzioni di disequazioni variazionali. Annali di Mat. Pura e Applicata **90**, 137–159 (1976)
2. Cioranescu, D., Donato, P.: An Introduction to Homogenization. Oxford University Press, Oxford (1999)
3. Cioranescu, D., Saint Jean Paulin, J.: Homogenization of Reticulated Structures. Springer, New York (1999)
4. Cioranescu, D., Donato, P., Zaki, R.: Periodic unfolding and Robin problems in perforated domains. CR Acad. Sci. Paris **7**, 467–474 (2006)
5. Cioranescu, D., Damlamian, A., Griso, G.: The periodic unfolding method in homogenization. SIAM J. Math. Anal. **40**, 1585–1620 (2008)
6. Cioranescu, D., Damlamian, A., Donato, P., Griso, G., Zaki, R.: The periodic unfolding method in domains with holes. SIAM J. Math. Anal. **44**, 718–760 (2012)
7. Cioranescu, D., Damlamian, A., Griso, G.: The Periodic Unfolding Method in Homogenization. Springer, Singapore (2018).
8. Courant, R., Hilbert, D.: Methods of Mathematical Physics. Interscience Publishers, New York (1962)
9. Donato, P., Jose, E.: Corrector results for a parabolic problem with a memory effect. ESAIM: Math. Model. Numer. Anal. **44**, 421–454 (2010)
10. Donato, P., Le Nguyen, K.H.: Homogenization of diffusion problems with a nonlinear interfacial resistance. Nonlinear Diff. Equ. Appl. **22**, 1345–1380 (2015)
11. Donato, P., Monsurró, S.: Homogenization of two heat conductors with an interfacial contact resistance. Anal. Appl. **2**, 1–27 (2004)
12. Donato, P., Le Nguyen, K.H., Tardieu, R.: The periodic unfolding method for a class of imperfect transmission problems. J. Math. Sci. **176**, 891–927 (2011)
13. Gemida, E.: Homogenization of an eigenvalue problem in a two-component domain with an interfacial barrier. Master's thesis, University of the Philippines- Los Baños (2018)
14. Hummel, H.K.: Homogenization for heat transfer in polycrystals with interfacial resistances. Appl. Anal. **75**(3–4), 403–424 (2000)
15. Kesavan, S.: Homogenization of elliptic eigenvalue problems. Appl. Math. Optim. **5**, Part I, 153–167, Part II, 197–216 (1979)
16. Monsurró, S.: Homogenization of a two-component composite with interfacial thermal barrier. Adv. Math. Sci. Appl. **13**, 43–63 (2003)
17. Monsurró, S.: Erratum for the paper Homogenization of a two-component composite with interfacial thermal barrier. Adv. Math. Sci. Appl. **14**, 375–377 (2004)
18. Vanninathan, M.: Homogenization of eigenvalue problems in perforated domains. Proc. Indian Acad. Sci. **90**, 239–271 (1981)
19. Zeidler, E.: Applied Functional Analysis: Applications to Mathematical Physics (Part I and Part II). Springer, New York (2012)

Printed in the United States
by Baker & Taylor Publisher Services